MAPPING RICE IN VIET NAM

ESTIMATING RICE AREA USING REMOTE SENSING AND AREA FRAME SURVEY METHODS IN AN GIANG PROVINCE

Anthony Burgard; Arturo Y. Pacificador, Jr.; Takaaki Masaki;
Anna Christine Durante; and Pamela Lapitan

OCTOBER 2025

ASIAN DEVELOPMENT BANK

6 ADB Avenue, Mandaluyong City, 1550 Metro Manila, Philippines
Tel +63 2 8632 4444; Fax +63 2 8636 2444
www.adb.org

ISBN 978-92-9277-470-7 (print); 978-92-9277-471-4 (PDF); 978-92-9277-472-1 (e-book)
Publication Stock No. TCS250416-2
DOI: http://dx.doi.org/10.22617/TCS250416-2

Note:
In this publication, "$" refers to United States dollars.
ADB recognizes "Korea" as the Republic of Korea and "Vietnam" as Viet Nam.

Cover design by Editha Creus. Featured on the cover, starting from left to right, are images capturing the fieldwork and remote sensing training session in Viet Nam (photo by JFPR TA 6856 Project Team).

Contents

Tables and Figures

Tables

Figures

Foreword

This study *Mapping Rice in Viet Nam: Estimating Rice Area Using Remote Sensing and Area Frame Survey Methods in An Giang Province* demonstrates a scalable approach for mapping rice cultivation in An Giang Province using satellite remote sensing and probability-based ground truth sampling, producing highly accurate rice area estimates that align closely with official statistics.

This study is a collaboration between the Asian Development Bank (ADB) and the National Statistics Office of Viet Nam (NSO), conducted under the Japan Fund for Prosperous and Resilient Asia and the Pacific as part of Technical Assistance (TA) 6856: Development of New Statistical Resources and Building Capacity in New Data Sources and Technologies. The TA project supports evidence-based policymaking by developing statistical frameworks that measure and monitor key sectors of the economy. It also aims to strengthen national statistical systems and enhance policy development by improving the technical capacity to produce relevant, accurate, and timely statistics. The study specifically aligns with Output 2 of the TA, which focuses on creating crop maps using digital technology and remote sensing.

The ADB project team extends sincere appreciation to the implementing agencies from the NSO and the Ministry of Agriculture and Environment for their essential contributions and valuable insights into agricultural practices throughout the Asia and Pacific region. The team is grateful to Mr. Le Trung Hieu, Deputy Director General, and Ms. Do Thi Thu Ha, Deputy Director of the Agriculture, Forestry, and Fishery Statistics Department from the NSO, who coordinated the implementation in the pilot areas.

The team acknowledges the technical expertise and collaboration of Mr. Sobue Shinichi from Japan Aerospace Exploration Agency, along with Ms. Pegah Hashemvand Khiabani and Mr. Naoki Furuta from the Remote Sensing Technology Center of Japan, whose technical contributions were instrumental to the project's success. We also extend our gratitude to the field and technical staff whose meticulous work in data collection, processing, and analysis formed the foundation of this study. Without the collective efforts of these individuals and organizations, this study would not have been possible.

The report has been produced by the Data Division within the Economic Research and Development Impact Department at ADB, under the overall direction of Elaine S. Tan. The project and report teams were led by Takaaki Masaki and Mahinthan Joseph Mariasingham. The core author team comprised Anthony Burgard; Arturo Y. Pacificador, Jr.; Anna Christine Durante; and Pamela Lapitan. Toby Miller copyedited the final manuscript, Nonie Villanueva typeset the report, and Editha Creus prepared the cover design.

It is hoped that this study contributes to advancing satellite-based methodologies for crop area statistics, potentially improving the timeliness, cost-effectiveness, and accuracy of agricultural statistics across developing countries in Asia and the Pacific. It is anticipated that this report will have a positive influence on the future of agricultural statistics.

Albert Francis Park
Chief Economist and Director General
Economic Research and Development Impact Department
Asian Development Bank

Introduction

Accurate and timely mapping of rice cultivation areas enables effective monitoring of agricultural production, allowing national governments to observe and respond promptly to changes in production and potential climate events or disasters due to natural hazards. Effective monitoring through mapping further supports reliable agricultural statistics, which are required for data-driven economic planning, food security, and policymaking. This is critical for Viet Nam, where rice holds economic importance both as the primary staple food and as a key export commodity. In 2023, Viet Nam produced 43.5 million metric tons of paddy rice cultivated across 7.12 million hectares. Domestic consumption accounted for 21.6 million metric tons, while exports reached 8.3 million metric tons, positioning Viet Nam as the world's third-largest rice exporter. Moreover, agriculture remains a key contributor to the real economy, employing approximately 32.9% of the country's labor force.[1]

An Giang Province,[2] one of Viet Nam's leading rice-producing regions and the focal area of this study, is an ideal region to implement rice mapping. With 616,200 hectares dedicated to rice cultivation and ranking second nationally in paddy production, the province faces significant environmental and disaster-related challenges. These include land subsidence, saltwater intrusion, and vulnerability to flooding and drought. Such risks highlight the need for accurate and regular monitoring of rice-growing areas to inform policy decisions and promote sustainable agricultural practices.

---

1 National Statistics Office of Viet Nam (NSO). 2024. *Statistical Yearbook of 2023.* https://www.nso.gov.vn/en/default/2024/07/statistical-yearbook-of-2023/.

2 The analysis presented in this report is based on the administrative boundaries of An Giang Province prior to its recent merger in June 2025, and does not reflect the newly configured provincial boundaries.

In response to these challenges, this study presents a methodology aimed at producing accurate rice cultivation maps for An Giang Province during the 2024 summer–autumn rice crop season. Our approach integrates remote sensing methods using publicly available satellite imagery, specifically through the Japan Aerospace Exploration Agency (JAXA)'s International Asian Harvest mOnitoring system for Rice (INAHOR), with a systematic framework for ground truth data collection. Implemented on the Google Earth Engine (GEE) platform, INAHOR integrates Synthetic Aperture Radar (SAR) data from JAXA's Advanced Land Observing Satellite-2 (ALOS-2) and optical imagery from the European Space Agency (ESA)'s Sentinel-2 constellation to estimate rice-growing areas.

Our approach relies on the adoption of a robust, grid-based area sampling methodology for collecting ground truth data, which informs the classification models, such as INAHOR, for the estimation of rice-planted area. This design builds on conventional sampling methods in ensuring unbiased selection of observation points that accurately reflect the observed variability across land classes.[3] We validate our rice area estimates by comparing them against official administrative data from the National Statistics Office of Viet Nam (NSO) to better understand the inherent biases from differing rice area estimation methods.

The methodology developed through this study seeks to establish a reproducible and scalable framework that extends its applicability beyond An Giang Province. It offers national statistical offices and agricultural data producers throughout the Asia and Pacific region a structured method for incorporating earth observation data and machine learning methods to produce crop area statistics. Ultimately, the introduction of remote sensing methods with a probability-based ground truth survey aims to enhance the accuracy, reliability, and operational efficiency of rice area monitoring in the Asia and Pacific region.

Current Rice Area Estimation Methods in Viet Nam

The NSO currently operates an administrative reporting system to estimate rice cultivation areas at commune levels throughout Viet Nam. The administrative reporting system covers all rural communes involved in crop cultivation and urban wards with at least 5 hectares under cultivation. Local officials, typically village chiefs or local government staff, serve as enumerators, compiling information through estimates taken during monthly field observations.[4]

The NSO applies two methodological approaches for estimating cultivated rice areas. The first, known as the Direct Aggregation Method, involves officials conducting field visits to assess crop conditions, estimate the planted areas by eye observation, and subsequently aggregate these

3 R. Congalton. 1988a. A Comparison of Sampling Schemes Used in Generating Error Matrices for Assessing the Accuracy of Maps Generated from Remotely Sensed Data. *Photogrammetric Engineering and Remote Sensing*. 54 (5). pp. 593–600; and R. Congalton. 1988b. Using Spatial Autocorrelation Analysis to Explore the Errors in Maps Generated from Remotely Sensed Data. *Photogrammetric Engineering and Remote Sensing*. 54(5). pp. 587–592.

4 National Statistics Office of Viet Nam (NSO). 2024. *Statistical Yearbook of 2023*. https://www.nso.gov.vn/en/default/2024/07/statistical-yearbook-of-2023/.

observations to produce total area estimates. The alternative method, known as the Exclusion Aggregation Method, is primarily used for single-crop fields. It calculates the cultivated area by deducting estimated unplanted areas from the total agricultural land.

Despite its comprehensive geographic coverage, this administrative reporting system encounters several limitations. One major issue is the inherent subjectivity and observer bias associated with local officials' self-reporting, which can result in inconsistent estimations across different regions and time periods. Maintaining the administrative reporting system requires extensive fieldwork, which poses logistical challenges, demands significant labor, and can be delayed by poor weather conditions limiting access to observation areas.

To mitigate these limitations and improve robustness of crop area statistics, there is a growing need to use more objective and precise measurement methodologies. Although global positioning system (GPS)-based area measurement surveys may be considered the gold standard for their accuracy and resource efficiency, they still face operational hurdles when implemented at a large national scale and may become cost-prohibitive when higher-frequency reporting is required.[5] Remote sensing technology offers a complement to traditional area measurement surveys, requiring potentially fewer field resources and improved timeliness, and serves as a complementary validation method to assess the quality and accuracy of compiled area statistics, rather than a complete replacement.

Moreover, remote sensing products can play a role in the traditional survey cycle by informing sample design and thus improving its efficiency, enhancing estimation methods through model-assisted estimators, and enabling more granular small-area estimates.

Advances in Remote Sensing for Agriculture

Remote sensing methodologies have become increasingly accessible to agricultural statisticians due to advancements in satellite data availability, analytical tools, and investments by official producers of agricultural data. First, satellite imagery is now more widely accessible online as a global public good through initiatives of national and regional space programs such as ESA's Sentinel missions, JAXA's ALOS-2 satellite data, and the National Aeronautics and Space Administration (NASA)'s Landsat program. Second, analytical tools for crop mapping such as JAXA's INAHOR and ESA Sen4Stat have simplified workflows, making remote sensing more accessible to non-remote sensing specialists for producing official statistics. Finally, national statistical offices and agricultural data producers have expanded their investments in geospatial data collection and analysis with the establishment of departments dedicated to using geospatial data.

5 C. Carletto et al. 2016. *Cheaper, Faster, and More than Good Enough: Is GPS the New Gold Standard in Land Area Measurement ?*. World Bank. https://documents.worldbank.org/en/publication/documents-reports/documentdetail/257481469116364362/Cheaper-faster-and-more-than-good-enough-is-GPS-the-new-gold-standard-in-land-area-measurement; and A. Burgard et al. 2024. *Plotting from Above: Enhancing Agricultural Mapping in Asia and the Pacific*. https://www.adb.org/sites/default/files/publication/976906/plotting-above-agricultural-mapping-asia-pacific.pdf.

JAXA's INAHOR, operational since 2013, has been used extensively across Southeast Asia for rice area estimation using SAR imagery from JAXA's ALOS-2 satellite. The INAHOR system was piloted under the Asian Development Bank (ADB) Technical Assistance Project (TA-8369) through pilot studies in Ang Thong (Thailand), Savannakhet (Lao People's Democratic Republic), Nueva Ecija (Philippines), and Thai Binh (Viet Nam).[6] Since these pilots were developed, JAXA and the Remote Sensing Technology Center of Japan have worked to simplify workflows by shifting compute-intensive calculations to the cloud, specifically the GEE platform. Previously, a major bottleneck was the need to download large image files and process them on local computers, requiring significant computing infrastructure. With cloud-based processing, users can now access satellite imagery directly from the cloud and perform complex calculations remotely, reducing computational burdens and improving timeliness.[7]

Similarly, ESA's platforms such as Sen4Stat (formerly Sen2Agri) help to improve regular monitoring of agricultural lands and key crops using satellite imagery from the Sentinel constellation of satellites.[8] Like JAXA's INAHOR system, Sen4Stat improves public access to Sentinel imagery, enables monitoring land classes over time, and outputs key statistics derived from agriculture maps. Notable applications include national-scale crop mapping initiatives in Ukraine and South Africa, as well as Senegal and Lesotho.[9]

At the national level, remote sensing technologies are widely utilized across the Asia and Pacific region for agricultural monitoring. Statistics Indonesia integrates ground truth data from its rice area survey with Sentinel-1 and Sentinel-2 satellite imagery to generate remote sensing-based rice area estimates.[10] In the Republic of Korea, the national Smart Farm Map system combines administrative data and remote sensing to produce key agricultural indicators across high-value crops.[11] In Japan, the Ministry of Agriculture, Forestry, and Fisheries employs remote sensing technologies to monitor agriculture, forest, and fisheries resources.[12] In Thailand, the Geo-Informatics and Space Technology Development Agency generates weekly crop estimates, including rice and other key economic crops, using its GISagro 4.0 platform and satellite data from Thaichote.[13] In the Philippines, the Philippine Rice Information System uses SAR imagery to estimate both rice area and yield.[14]

---

6 A.C. Durante et al. 2018. Improving Paddy Rice Statistics Using Area Sampling Frame Technique. *ADB Economics Working Paper Series*. No. 565. Asian Development Bank.

7 Japan Aerospace Exploration Agency (JAXA). 2023. *High-Resolution Land-Use and Land-Cover Map of Vietnam for 2020* (Version 23.09). https://www.eorc.jaxa.jp/ALOS/en/dataset/lulc/lulc_vnm_v2309_e.htm.

8 European Space Agency (ESA). 2023. *Sentinels for Agricultural Statistics – Final Report*. https://eo4society.esa.int/wp-content/uploads/2023/02/esa-sentinels-for-agricultural-statistics.pdf.

9 P. Defourny et al. 2019. Near Real-Time Agriculture Monitoring at National Scale at Parcel Resolution: Performance Assessment of the Sen2-Agri Automated System in Various Cropping Systems around the World. *Remote Sensing of Environment*. 221. pp. 551–568; Food and Agriculture Organization (FAO). 2018. *EOSTAT: Trusted Methods for the Use of EO Data for Crop Type Mapping and Crop Yield Forecasting*. https://openknowledge.fao.org/bitstreams/32767090-b0be-40a2-9831-c1928fd13a6d/download; and F. J. Gallego. 2004. Remote Sensing and Land Cover Area Estimation. *International Journal of Remote Sensing*. 25 (15). pp. 3019–3047.

10 L. Gandharum et al. 2021. Remote Sensing Versus the Area Sampling Frame Method in Paddy Rice Acreage Estimation in Indramayu Regency, West Java Province, Indonesia. *International Journal of Remote Sensing*. 42 (5). pp. 1738–1767.

11 S. Wee. 2024. *Improvement and Utilization of Farm Map (Agricultural Digital Map)*.

12 Remote Sensing Technology Center of Japan (RESTEC). 2024. https://www.restec.or.jp/en/service/agriculture/index.html.

13 Geo-Informatics and Space Technology Development Agency (GISTDA). 2025. https://www.gistda.or.th/home.php.

14 Philippine Rice Information System (PRiSM) 2025. Data Products. https://prism.philrice.gov.ph/ (accessed 11 June 2025).

Remote Sensing Estimates Depend on High-Quality Ground Truth Data

Producing accurate crop maps with remote sensing relies on high-quality ground truth data, which serves as a reference for machine learning models to perform land cover classification. These models use ground truth data to interpret satellite imagery and differentiate between various land cover and crop types. For instance, to distinguish rice fields from other crops, models analyze spectral indices derived from satellite imagery. These indices capture differences in vegetation reflectance, allowing models to recognize unique patterns associated with specific crop types. These patterns are akin to a fingerprint that uniquely provides discriminating power to classify a pixel or object as a particular land use or crop. The ability to discriminate between land use and crop classes varies widely between the type of crops and is an area of productive and evolving research to better understand the spectral patterns of local crops in different agroecological zones. Rice, for example, is usually identified by its distinct phenological development. It begins its growth in flooded fields, greens as it matures, and turns yellow at harvest. These unique characteristics allow us to distinguish rice not by spectral indices at a single point in time but by tracking changes in values throughout the growing season.

Each ground truth data observation consists of three components: a precise geographic location, the time of observation, and a label of observed characteristics. Precise location data allow a remote sensing specialist to match observed labels with the corresponding satellite imagery, while accurate timing ensures that the label reflects the appropriate crop season and phenological stage.

Ground truth data can be collected through field surveys or derived from high-resolution satellite imagery. The most accurate method involves field surveys, where field staff with subject matter expertise physically visit observation points, record exact geographic locations, and document the specific land cover or crop type observed. This method provides the most accurate validation of actual ground conditions, often including photos for further investigation for validation.

Alternatively, ground truth data can sometimes be generated through manual interpretation of high-resolution satellite imagery. In this approach, subject-matter experts visually analyze, and label different land cover types based on image characteristics. However, this method has limitations, especially when using lower-resolution or outdated imagery. Even with skilled experts, visual interpretation of low-resolution images can introduce labeling biases. In some cases, the imagery may lack sufficient detail to determine the correct label or may represent an unknown point in time, reducing its reliability for accurate classification.

Sample Size Determination for Ground Truth Data

Collecting an appropriate number of high-quality ground truth samples is essential to ensure a highly accurate crop map. The required sample size depends on the number of classes targeted, the similarity and rarity of classes, the complexity of the study area, the quality and number of satellite images (and spectral bands) available during the reference period, and the relative accuracy of the mapping output.[15] For example, a simple binary classification (such as rice vs. non-rice) will require fewer samples compared to a larger classification scheme, where classes may be spectrally similar.[16] Targeting rarer classes likewise necessitates more samples to account for the smaller proportion, and more targeted stratified sampling methods are essential to ensure an adequate representation of that class.[17]

The complexity of the study area is another factor influencing sample size. Homogenous agricultural areas with large monocropped parcels require fewer samples to capture variability. Areas with mixed cropping patterns, small field sizes, and diverse terrain would require a greater number of samples to account for the heterogeneity.

The availability of high-quality satellite imagery, in terms of its spatial, temporal, and spectral resolution further influences the sample size for crop mapping. High-resolution, multi-temporal datasets such as JAXA's ALOS-2 and ESA Sentinel-2 can provide detailed and diverse spectral information, improving discrimination of spectrally similar crop types. High spatial resolution imagery is particularly beneficial in fragmented agricultural landscapes, as it reduces the occurrence of mixed pixels, where a single pixel may contain multiple classes.[18] Conversely, imagery with lower and spectral resolution, or with infrequent temporal coverage, especially in areas with high cloud cover, offers less discriminative power and would thus require additional samples to compensate.

The purpose and accuracy of the mapping output should also be considered. Here, we determine accuracy by how well the mapped classifications match the actual observed classes in the ground truth data. Many studies base their assumptions on probability sampling theory, utilizing sample size calculations based on the binomial distribution and optimizing for a level statistical precision in the overall accuracy measures of the mapping product.[19] Samples are

---

15 H. Zhang et al. 2023. Comparing Three Methods of Selecting Training Samples in Supervised Classification of Multispectral Remote Sensing Images. *Sensors*. 23 (20). 8530; and S. Shetty et al. 2021. Assessing the Effect of Training Sampling Design on the Performance of Machine Learning Classifiers for Land Cover Mapping Using Multi-Temporal Remote Sensing Data and Google Earth Engine. *Remote Sensing*. 13 (8). 1433.

16 D. Moraes et al. 2021. Influence of Sample Size in Land Cover Classification Accuracy Using Random Forest and Sentinel-2 Data in Portugal. *2021 IEEE International Geoscience and Remote Sensing Symposium IGARSS*. Brussels: IEEE.

17 S. Stehman. 2009. Sampling Design for Accuracy Assessment of Land Cover. *International Journal of Remote Sensing*. 30. pp. 5243–5272.

18 M. Belgiu, and O. Csillik. 2018. Sentinel-2 Cropland Mapping Using Pixel-Based and Object-Based Time-Weighted Dynamic Time Warping Analysis. *Remote Sensing of Environment*. 204. pp. 509–523.

19 K. Fitzpatrick-Lins. 1981. Comparison of Sampling Procedures and Data Analysis for a Land-Use and Land-Cover Map. *Photogrammetric Engineering and Remote Sensing*. 47 (3). 343–351.

typically allocated in proportion to the relative area of each land cover class of interest, with a minimum number of samples per class to ensure a representative distribution of spectral values for accurate classification.

Congalton (1991) expands on the concept of recommended sampling for ground truth data, emphasizing that sample size should be determined by both the number of map classes and the study area's size.[20] A widely accepted rule of thumb suggests a minimum of 50 samples per class, but for larger study areas or maps with many classes, this could be increased to 75 or 100 samples per class. Studies further highlight the need to consider the allocation of ground truth samples to avoid issues of spatial auto-correlation bias. Practices here include the introduction of additional strata, i.e., agroecological zones, administrative areas, to improve sampling efficiency, and constructing area sampling frames based on non-overlapping segments for selection and ultimately observation.

WorldCereal (2025) and Gandharum et al. (2021) further propose a multi-stage sampling design using a grid-based area frame sampling approach. By dividing the study area into a series of nested spatial segments (squares or hexagons) and overlaying them with a previous land cover map, an assessment of spatial variability can be determined.[21]

This approach enables a more informed allocation of ground truth samples. In WorldCereal, this approach takes on a three-stage hexagon design, where the mixture of land classes or crops within segments are determined. Samples are then allocated in areas with known classes, enhancing the efficiency and precision of the design. Gandharum et al. rely on a similar nested square grid structure based on the assessment of similar Sentinel-1 spectral indices to cluster areas based on spectral similarity.[22]

Crop monitoring systems such as JAXA's INAHOR and ESA's Sen4Stat rely on supervised learning models for crop classification, leveraging enough high-quality ground truth data. Among these models, Random Forest remains the most widely used due to its versatility and efficiency. While alternative models like support vector machines, gradient boosting, and deep learning approaches such as convolutional and recurrent neural networks are also utilized in the literature, Random Forest continues to be the workhorse of these classification systems, renowned for its relative accuracy and processing speed.[23] First, it handles high-dimensional and multisource data well, incorporating features from optical satellite imagery, such as Sentinel-2, and radar SAR imagery, such as ALOS-2. The ability to fuse multisource imagery is particularly helpful for rice, providing greater flexibility in tropical climates where cloud cover may be a concern. Second, Random Forest performs well in data-scarce environments, making it suitable for large-scale applications where ground truth data may be limited. Finally, Random Forest integrates seamlessly with cloud-based processing platforms such as GEE.

---

20 R. Congalton. 1991. A Review of Assessing the Accuracy of Classification of Remotely Sensed Data. *Remote Sensing of Environment*. 37 (1). pp. 35–46.

21 L. Gandharum et al. 2021. Remote Sensing Versus the Area Sampling Frame Method in Paddy Rice Acreage Estimation in Indramayu Regency, West Java Province, Indonesia. *International Journal of Remote Sensing*. 42 (5). pp. 1738–1767; and WorldCereal. 2025. https://esa-worldcereal.org/en.

22 Ibid.

23 M. Belgiu, and O. Csillik. 2018. Sentinel-2 Cropland Mapping Using Pixel-Based and Object-Based Time-Weighted Dynamic Time Warping Analysis. *Remote Sensing of Environment*. 204. pp. 509–523.

Methodology

Objectives

This study aims to develop and validate a methodology for estimating rice cultivation areas using remote sensing in An Giang Province, Viet Nam, conducted during the 2024 summer–autumn rice cropping season. Specifically, the study seeks to establish a systematic, grid-based area frame sampling framework, combined with JAXA's INAHOR remote sensing platform, to produce accurate rice cultivation maps. The grid-based sampling framework, aligned with WorldCereal (2025) and Gandharum et al. (2021), is proposed to optimize the allocation of ground truth data collection points, thereby enabling thorough validation of remote sensing-derived classifications through field observations (footnote 21).

To assess the accuracy and statistical validity of rice area estimates, this study compares results from three distinct approaches: area estimates collected during ground truth surveys, INAHOR remote sensing classifications calibrated with ground truth data, and official administrative statistics from NSO. This comparative analysis aims to identify each method's relative strengths, limitations, and inherent biases in estimating rice cultivation areas.

Study Area

The geographical scope of this study covers the official administrative boundary of An Giang Province, situated in the southwestern region of Viet Nam's Mekong Delta, covering approximately 3,500 square kilometers (km^2) with a population of 2.2 million. The province shares a border with Cambodia to the northwest, and Dong Thap, Kien Giang, and Can Tho provinces to the east and south. The province is geographically defined by two major branches of the Mekong River: the Hậu Giang and Tiến Giang. Topographically, the province is predominantly characterized by flat terrain, except for the elevated Cấm Mountains in its western area. An Giang Province also hosts an extensive network of canals and rivers, which are the source for the natural irrigation of rice cultivation areas.

An Giang Province experiences a tropical monsoon climate with two primary seasons: a rainy season from May to November and a dry season from December to April, with an average temperature of approximately 27°C and high humidity. The province's agricultural crop calendar includes three notable rice-growing seasons: Winter-Spring, Summer-Autumn, and Autumn-Winter. The Winter-Spring season involves sowing in November–December and harvesting in March–April; the Summer-Autumn season entails sowing in April–May and harvesting in July–August; and the Autumn-Winter season comprises sowing between July and September, with harvesting from October to December. Among these, the Winter-Spring and Summer-Autumn seasons are the major rice-producing periods. However, not all farmers participate in all three cycles; some opt to rotate crops, such as cultivating lotus flowers, to help replenish the soil. Recent agricultural trends have seen a turnover of rice growing areas to higher value-added crops, such as durian fruit, particularly in Cho Moi district.

The mapping conducted in this study corresponds to the Summer-Autumn rice season with ground truth observations taken in August 2024, aligning with the harvest period of the Summer-Autumn rice-growing season. This seasonal cycle typically starts with sowing or transplanting activities in April-May and culminates with harvest from July to August.[24] While efforts were taken to observe rice during the harvest season, it is noted that actual rice observed was at asynchronous stages of growth: some plots were just flowering, others were ready for harvest, and certain areas had already been harvested.

Data Sources

Satellite imagery

The study utilizes two satellite imagery products through the GEE catalog. The first product, PALSAR-2 ScanSAR Level 2.2 (JAXA/ALOS/PALSAR-2/Leve2_2/ScanSAR), is SAR data provided by JAXA through the Earth Observation Research Center. This product is available at a spatial resolution of 25 meters, and contains horizontal-horizontal (HH) and horizontal vertical (HV) polarizations, local incidence angle, and a data quality mask band.

24 *Remote Sensing Methodology Training*. Ministry of Agriculture and Rural Development, 10 May 2025.

The second product, Harmonized Sentinel-2 Multispectral Instrument (COPERNICUS/S2_SR_HARMONIZED), is optical imagery provided through the Copernicus program of the ESA. The product is available at a spatial resolution of 10–60 meters, depending on the band. In total, Sentinel-2 includes 13 spectral bands spanning from the visible to shortwave infrared spectrum. Imagery for both products were acquired for the study for the duration of the Summer-Autumn rice season from April 2024 to August 2024.

A derived High-Resolution Land-Use and Land-Cover (LULC) Map of Viet Nam for 2020 from JAXA, cropped to An Giang Province, was also utilized to build the sampling frame. The products specify 12 land use and cover classes, including a rice class (footnote 7). The total area of the province is estimated at 3,616.24 km^2, of which 2,557.03 km^2 are classified as rice-planted areas (70.71%).

INAHOR further utilizes ESA's WorldCover and NASA's Moderate Resolution Imaging Spectroradiometer (MODIS) products as part of its post-processing.

Ground truth data

ADB and NSO conducted a field survey to collect ground truth data for classifying rice areas using the INAHOR system from 13 to 16 August 2024. A total of 272 observation points were recorded across 34 grids. Each observation point covered an area of 50 m x 50 m. Field staff successfully observed all planned points during the 4-day data collection exercise.

The median distance between the allocated sampling point and the actual observation point was 59 meters, demonstrating a strong alignment between planned and field-observed locations. This close alignment is a key data quality indicator, as minimal deviation from the planned sampling points helps reduce sampling bias. Specifically, the median distances by land use are as follows: 61 meters for Rice, 27 meters for Other Crops, 19 meters for Urban, and 429 meters for Water; the latter being higher because most observations were made from land overlooking water bodies.

These deviations occurred primarily due to difficulties navigating agricultural areas without damaging crops. In the extensive monocropped rice paddy areas of An Giang Province, enumerators successfully identified the correct crop from a distance. During data processing, standing locations were adjusted to the center of each parcel based on verification of field photographs. This cleaning adjustment relocated location points closer to parcel centers to avoid potential bias from edge areas where spectral values may not be representative of the overall parcel characteristics. At each observation point, field teams recorded GPS coordinates using consumer-grade tablets, classified the observed land cover according to a binary classification scheme (rice and non-rice), estimated the percentage of rice within the 50 m by 50 m segment, and captured photos from the point of observation in four cardinal directions. Data collection was facilitated by the World Bank's Survey Solutions software.[25] The full questionnaire used in the data collection is provided in Appendix 2. Ground Truth Survey Questionnaire Form.

25 World Bank. 2025. Survey Solutions CAPI/CAWI platform: Release 25.01.2. Washington DC, The World Bank.

Administrative data

For external validation of the study results, the NSO provided administrative data on the area planted with rice. The dataset includes the total rice-planted area, measured in hectares, for the 2024 Summer-Autumn season. Data are available at both the province level and for each of the 11 districts within the province.

Sampling Frame and Design

The sampling frame for this study is based on JAXA's High-Resolution LULC Map of Viet Nam for 2020 (footnote 7). The LULC originally detailed 12 land-use and -cover classes at a 10-meter resolution, with classification accuracy of the rice class exceeding 90%. For sample design and the formation of sampling strata of this study, the 12 land-use and -cover classes were consolidated into four main classes: rice, other crops, built-up urban areas, and water bodies (Figure 1). This aligned with our categories of land cover classification in the field and the reduced number of classes helps to minimize any misclassification of non-rice areas. Under this reclassification, approximately 70.7% of the land area is under rice cultivation, while forested and urban areas, especially near the provincial capital, Long Xuyen, represent the next largest land use categories.

Figure 1: Reclassified Land Cover Map, Representing Rice, Other Crops, Urban, and Water

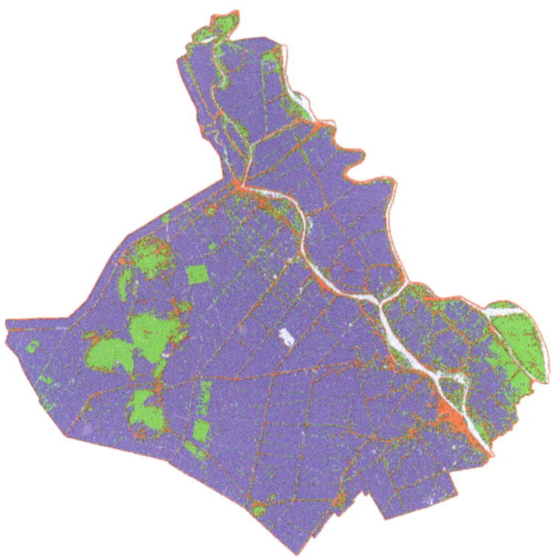

Re-classification mapping:
Class 1: Rice (Class 3)
Class 2: Other crops (Classes 4–11)
Class 3: Urban (Class 2)
Class 4: Water (Classes 1,12)

Note: Japan Aerospace Exploration Agency Land-Use and Land-Cover classification: (1) Water, (2) Urban/Built-up, (3) Rice, (4) Other Crops, (5) Grass/Shrub, (6) Woody Crops/Orchards, (7) Barren, (8) Evergreen Forest, (9) Deciduous Forest, (10) Plantation Forest, (11) Mangrove Forest, (12) Aquaculture. The map reflects the administrative boundaries of An Giang Province prior to its recent merger in June 2025. All classification and analysis are based on the former provincial configuration.

Source: Japan Aerospace Exploration Agency, Earth Observation Research Center (JAXA EORC). https://www.eorc.jaxa.jp/ALOS/en/dataset/lulc/lulc_vnm_v2309_e.htm; Asian Development Bank visualization using QGIS 3.34.14. 2024. Manila.

This study utilizes a three-stage square grid sampling design to balance representativeness and logistical efficiency during field operations. In the first stage, the study area is divided into primary sampling units consisting of 1 km x 1 km grids (GRIDs), from which a stratified random sample is selected. Each selected GRID is subsequently subdivided into 25 smaller sub-grids (SUBGRIDs) of 200 m x 200 m during the second stage. A serpentine systematic selection of these SUBGRIDs is applied to minimize spatial bias that could arise from selecting adjacent grids. In the third stage, each selected SUBGRID is further divided into 16 segments (SEGMENT) measuring 50 m x 50 m, from which SUB-SEGMENTs are also selected through serpentine systematic selection. We finally mark the centroid of each segment as the observation point. Figure 2 illustrates this nested, multi-stage sampling structure.

Figure 2: Example of 1 km x 1 km Nested Three-Stage Grid

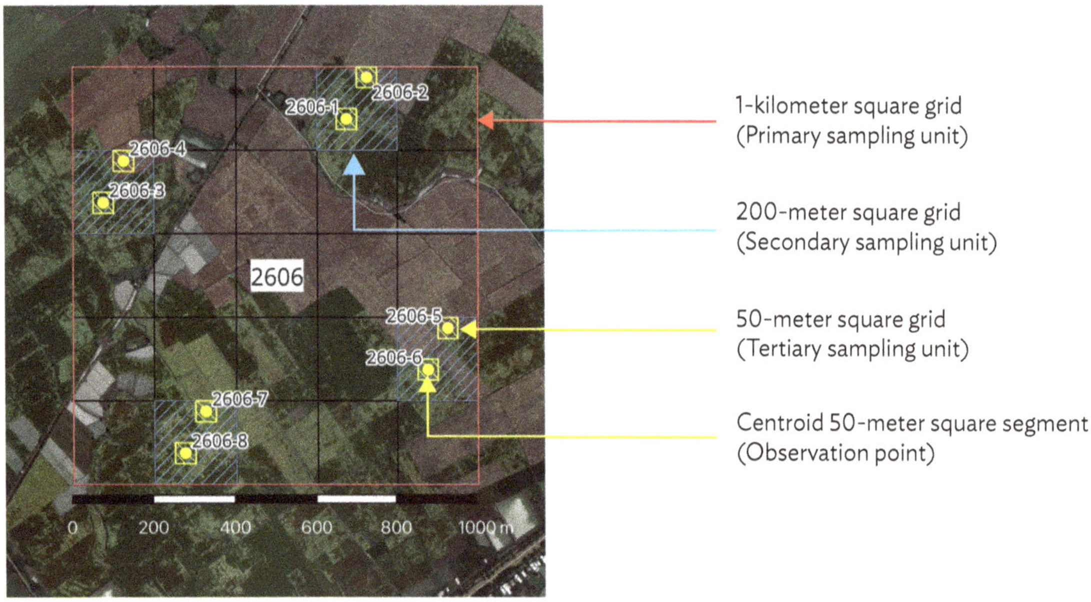

Source: Asian Development Bank visualization using QGIS 3.34.14. 2024. Manila.

A GRID was selected to optimize operational efficiency, as it is determined to be large enough to capture variability in land-use classes but small enough to minimize travel time during fieldwork and to maintain budgetary constraints. Within each GRID, a SUBGRID was established, roughly corresponding to the average size of a rice parcel. Finally, a SEGMENT was defined as the ultimate sampling unit. This SEGMENT matches the minimum mapping unit of the rice area map, intentionally capturing a cluster of pixels to better represent the observed class.

To enhance the representativeness and precision of the sample, stratification is applied. The GRIDs are divided into non-overlapping strata based on land use characteristics. Administrative districts form one stratum to ensure sample GRIDs are drawn from each of the 11 districts, allowing for district-level estimates of total rice-planted area. Other strata include GRIDs where more than 90% of the area is planted with crops other than rice, GRIDs where over 90% consists of built-up urban areas, and GRIDs where more than 90% of the area is covered by water bodies.

For each GRID, SUBGRID, and SEGMENT, the total number of 10 m × 10 m pixels from the reclassified land cover map are calculated, along with the number of pixels classified as rice, other crops, built-up urban areas, and water bodies. The count of pixels informs the variability in rice cultivation and other land use classes both within and between these spatial units.

Sample Size Determination

The sample size is determined to ensure a representative selection of segments that accurately reflect the characteristics of the (study area) population. In this context, the population refers to the features of the segments, and a good representation means choosing a sample size that enables an accurate estimate of the total area planted with rice, with a coefficient of variation (CV) below 10%.

To derive the required sample size that meets this criterion, we first examine the estimation of the total area planted with rice with respect to the three-stage sampling design. The total rice area is estimated by counting rice pixels within nested sampling units: GRIDs, SUBGRIDs, and SEGMENTs. The study area is first divided into 1 km x 1 km grids A, a, then into 200 m x 200 m sub-grids B,b, and finally into 50 m x 50 m segments D,d. Within each segment, rice pixels overlapping with the reclassified JAXA LULC are counted. The total area of a grid planted with rice (in square meters) is calculated as the product of the number of rice pixels and 100 m^2 (the area of a single pixel). The total rice area across the study area is then estimated as:

$$\hat{Y} = \frac{A}{a}\sum_{i=1}^{a} B_i \frac{1}{b_i}\sum_{j=1}^{b_i} D_{ij}\frac{1}{d_{ij}}\sum_{k=1}^{d_{ij}} y_{ijk} = \sum_{i=1}^{a}\sum_{j=1}^{b_i}\sum_{k=1}^{d_{ij}} w_{ijk}\, y_{ijk}. \quad (1)$$

where y_{ijk} is the number of rice pixels in the k-th segment of the j-th sub-grid in the i-th grid, and

$$w_{ijk} = \frac{A}{a} \times \frac{B_i}{b_i} \times \frac{D_{ij}}{d_{ij}} \quad (2)$$

is the expansion weight, which adjusts the observed counts to represent the total rice area. The theoretical variance of this estimator is decomposed into three components, capturing the variability at different sampling stages:

$$V(\hat{Y}) = \underbrace{A^2\left(\frac{1}{a}-\frac{1}{A}\right)S_1^2}_{\text{Between Grids Variance}} + \underbrace{\frac{A}{a}\sum_{i=1}^{A} B_i^2\left(\frac{1}{b_i}-\frac{1}{B_i}\right)S_{2i}^2}_{\text{Between Sub-grids Variance}} + \underbrace{\frac{A}{a}\sum_{i=1}^{A}\frac{B_i}{b_i}\sum_{j=1}^{B_i} D_{ij}^2\left(\frac{1}{d_{ij}}-\frac{1}{D_{ij}}\right)S_{ij}^2}_{\text{Between Segments Variance}} \quad (3)$$

Each component quantifies the contribution of variability at its respective stage of sampling. The between-grids variance dominates, indicating that differences in rice pixel counts across 1 km x 1 km grids are the largest contributors to overall uncertainty. To achieve an acceptable level of precision (CV < 10%) while maintaining budget constraints, sample sizes at the grid a, sub-grid b, and segment d levels can be varied to optimize their relative allocations based on variance contributions and cost considerations.

We further assume that the variability between the 1 km × 1 km grids is much larger than the variability within the grids, which allows us to simplify the overall variance estimator of the total rice area, $V\left(\hat{Y}\right)$, by focusing solely on the grid-level component. In this simplified approach, the variance of the total estimate is given by:

$$V\left(\hat{Y}\right) \approx A^2\left(\frac{1}{a} - \frac{1}{A}\right)S_1^2 \qquad (4)$$

where:

A is the total number of grids, and a is the number of grids sampled.
S_1^2 is the variance between the grid totals, computed as:

$$S_1^2 = \frac{1}{A-1}\sum_{i=1}^{A}\left(Y_{i..} - \overline{Y}^*\right)^2 \qquad (5)$$

with

$Y_{i..}$ representing the total count of rice pixels in grid i, summed over all its sub-grids and segments:

$$Y_{i..} = \sum_{j=1}^{B_i}\sum_{k=1}^{D_{ij}} y_{ijk} \text{ , grid totals} \qquad (6)$$

and $\overline{Y}^*$ is the overall mean of these grid totals:

$$\overline{Y}^* = \frac{1}{A}\sum_{i=1}^{A} Y_{i..} \qquad (7)$$

The term $\left(\frac{1}{a} - \frac{1}{A}\right)$ acts as a finite population correction, adjusting for sampling without replacement from a finite number of grids.

The CV for our estimator, defined as the ratio of its standard deviation to the mean, serves as the metric for determining the required sample size.

$$CV\left(\hat{Y}\right) = \frac{\sqrt{V\left(\hat{Y}\right)}}{\hat{Y}} \times 100\%. \qquad (8)$$

Based on the target CV of 10%, it was determined that drawing 34 sample grids is necessary to achieve a CV below the 10% threshold.

To enhance the representativeness of the sample, GRIDs were stratified to ensure coverage across all districts and key land types, specifically those primarily planted with other crops, urban/settlement areas, and water bodies. This stratification was designed to improve the classification performance.

A total of 34 GRIDs were allocated, with two sample GRIDs systematically selected per district after removing those where more than 90% of the area was covered by other crops, urban/settlements, or water bodies. The systematic selection ensured that the sampled GRIDs were spatially dispersed. Additionally, systematic samples were drawn from the different strata, including six GRIDs from 120 in the crop stratum, two from five in the urban stratum, and four from 25 in the water stratum.

In each sample GRID, a systematic sample of four SUBGRIDs is selected (= 4). Within each sample SUBGRID, a systematic sample of two segments is selected (= 2). More SUBGRIDs were chosen compared to segments to account for the relatively higher larger variation. This results in a total of 272 SEGMENTs. A minimum of two segments per SUBGRID is required to ensure a reliable variance estimation.

A basic principle in area frame sampling is to ensure that sample areas are not clustered too closely together. To achieve this, SUBGRIDs are arranged in a serpentine pattern, ensuring that when a systematic sample of four SUBGRIDs is selected, they are well distributed across the grid. For regularly sized GRIDs, Figure 2 illustrates the serpentine ordering of SUBGRIDs and the systematic selection process. A similar approach was applied to SEGMENTs, where they were also arranged in a serpentine order to ensure spatial dispersion. Figure 3 depicts the ordering of SEGMENTs (Panel A) and the systematic selection of two sample SEGMENTs (Panel B).

Figure 3: Serpentine Systematic Selection of Subgrids and Segments

A) Selection of Sample Subgrids

1	2	3	4	5
10	9	8	7	6
11	12	13	14	15
20	19	18	17	16
21	22	23	24	25

Sampling Interval: 25/4≅6
Random Start: 4

B) Selection of Sample Segments

1	2	3	4
8	7	6	5
12	11	10	9
13	14	15	16

Sampling Interval: 16/2=8
Random Start: 3

Source: Asian Development Bank visualization. 2024. Manila.

For irregular-shaped GRIDs, the same serpentine ordering principle is applied to ensure a well-distributed selection of sample SUBGRIDs. This approach is also used for selecting segments from irregular SUBGRIDs to maintain spatial dispersion. Figure 4 illustrates the ordering and selection of sub-grids within irregular GRIDs.

Figure 4: Selection of Subgrids from an Irregular Grid

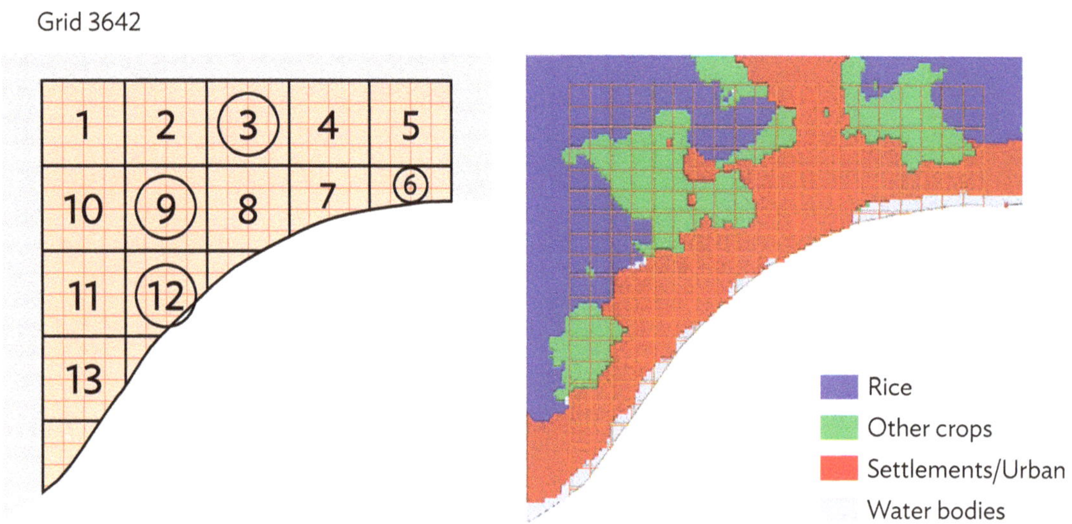

Source: Asian Development Bank visualization using QGIS 3.34.14. 2024. Manila.

Estimators of Total Area Planted with Rice

Several estimators for the total area planted with rice were considered, each offering an approach to handling the sample data and the available auxiliary information to improve the precision of the estimates. The first three estimators, the expansion estimator, the ratio estimator, and the adjusted ratio estimator excluding segments with matching issues, are based on the observed percentage of rice measured within 50 m × 50 m segments collected during the ground truth survey.

Expansion estimator

The expansion estimator is a design-based approach that directly scales up the percentage of rice observed in each 50 m × 50 m segment to estimate the total rice-planted area. This estimator relies on the known sampling probabilities of the segments, using sampling weights to scale the observed data to the total population area. Using the expansion estimator, both the estimate and its variance are calculated according to Equations (1) and (2).

Ratio estimator

The ratio estimator utilizes auxiliary information which leads to a more precise estimator especially when it is at least moderately correlated with observed area planted with rice. In this situation, the total area planted with rice from the area frame is constructed using 2020 JAXA LULC. The proposed ratio estimator is defined as:

$$\hat{Y}_{R1} = \frac{\hat{Y}}{\hat{X}}X = \frac{\sum_{i=1}^{a}\sum_{j=1}^{b_i}\sum_{k=1}^{d_{ij}} w_{ijk}y_{ijk}}{\sum_{i=1}^{a}\sum_{j=1}^{b_i}\sum_{k=1}^{d_{ij}} w_{ijk}x_{ijk}}X, \quad (9)$$

where the weight for each segment is given by

$$w_{ijk} = \frac{A}{a} \times \frac{B_i}{b_i} \times \frac{D_{ij}}{d_{ij}} \quad (10)$$

The variance of the ratio estimator is estimated as

$$s^2\left(\hat{Y}_{R1}\right) = s^2\left(\hat{Y}\right) + \hat{R}^2 s^2\left(\hat{X}\right) - 2\hat{R}s\left(\hat{Y},\hat{X}\right) \quad (11)$$

with

$$\hat{R} = \frac{\hat{Y}}{\hat{X}} \quad (12)$$

It must be noted that the ratio in (12) can be interpreted as a measure of the difference of total area planted with rice between the two years and should be considered in interpretation of its results.

Adjusted ratio estimator

Observations revealed that some segment-level characteristics, particularly the total area planted with rice, exhibited matching problems that resulted in large discrepancies, which could affect the precision of the ratio estimate. To address this issue, an adjusted ratio estimate was calculated by excluding 25 problematic segments as non-responses and adjusting the survey weights accordingly. Although this exclusion reduces the effective sample size, it minimizes bias in the estimation, ensuring that only reliable data contribute to the total estimate. The estimator and variance of the adjusted ratio estimator is the same as the ratio estimator detailed in equations 9–12.

Table 1 summarizes the key characteristics of three estimation methods: the Expansion Estimator, Ratio Estimator, and Ratio Estimator (Excluding Matching Issues). It highlights the main differences among these approaches across essential dimensions, including their use of auxiliary data, bias, variance, sample size requirements and data sensitivities.

Table 1: Comparison of Estimation Methods: Expansion, Ratio, and Adjusted Ratio

Estimators	Use of Auxiliary Data	Bias and Variance	Sample Size	Potential sensitivity to data issues
Expansion	None	Unbiased design, vulnerable to response or coverage bias; High variance	Baseline	Measurement errors and non-response
Ratio	JAXA LULC total rice area	Bias introduced but mitigated with large samples; Lower variance if strong correlation with auxiliary	Potentially more efficient is correlation with auxiliary	Quality and accuracy of the auxiliary data
Adjusted ratio	Uses auxiliary data to remove problematic segments before estimation	Risk of selection bias from excluded segments; lower variance by removal of outliers	Potentially more efficient	Rules for data exclusion

JAXA = Japan Aerospace Exploration Agency, LULC = High-Resolution Land-Use and Land-Cover Map.

Source: Asian Development Bank summary using data from ground truth survey and Japan Aerospace Exploration Agency. High-resolution land-use and land-cover map of Viet Nam for 2020 (version 23.09) (accessed August 2024).

Classifying rice area through INAHOR in Google Earth Engine

The JAXA INAHOR platform is an online tool built on GEE that generates rice area maps. It requires minimal input from the user: the study area's extent, field-obtained ground truth data provided as GPS point coordinates classified as rice or non-rice, and administrative boundaries to compile area statistics for each region. The acquisition and processing of satellite data is then handled by INAHOR based on the user inputs. The complete workflow for the INAHOR system is outlined in Figure 5. Further technical specifications of the INAHOR workflow are detailed in Appendix 3.

Figure 5: International Asian Harvest mOnitoring system for Rice Workflow from Input to Output

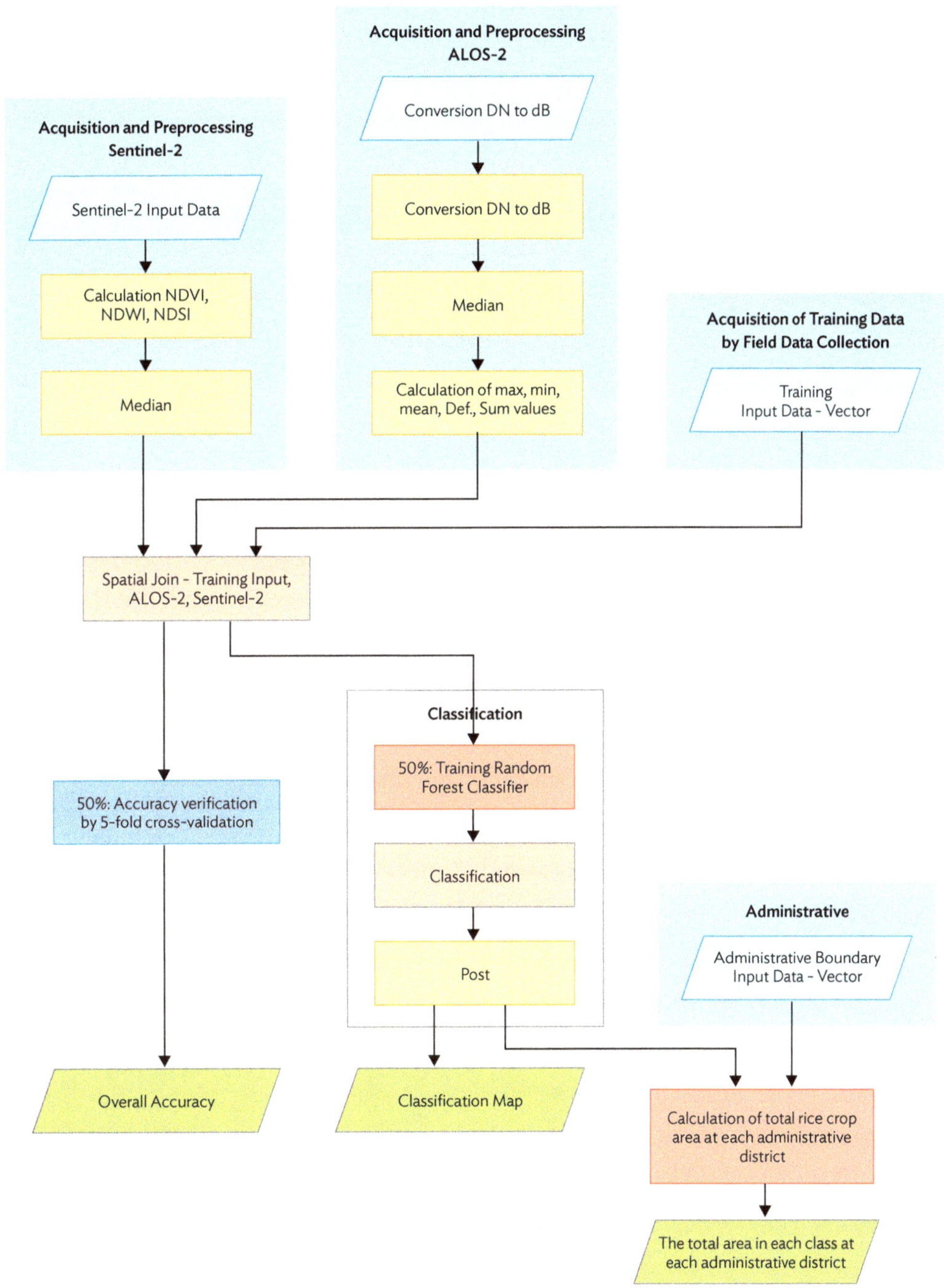

ALOS-2 = Advanced Land Observing Satellite-2, DN = digital number, dB = database, max = maximum, min = minimum, NDSI = Normalized Difference Snow Index, NDVI = Normalized Difference Vegetation Index, NDWI = Normalized Difference Water Index.

Source: Japan Aerospace Exploration Agency. International Asian Harvest mOnitoring system for Rice, 2024.

The core of the INAHOR platform is a Random Forest machine-learning classifier. In the context of rice area classification, Random Forest is an ensemble model that builds multiple decision trees using features derived from satellite imagery, such as spectral indices and radar backscatter values (Table 2). Each tree then independently classifies pixels as rice or non-rice, and the final classification is determined by aggregating these predictions, typically through majority voting.

Table 2: Spectral Features Used by International Asian Harvest mOnitoring system for Rice in the Random Forest Classifier

Type	Features	Source
Optical	Normalized Difference Vegetation Index Normalized Difference Water Index Normalized Difference Snow Index	ESA Sentinel-2
SAR	Horizontal-Horizontal (HH)-polarization, Horizontal-Vertical (HV)-polarization, Sum of HH and HV, and Difference of HH and HV	JAXA ALOS-2

ALOS-2 = Advanced Land Observing Satellite-2, ESA = European Space Agency, JAXA = Japan Aerospace Exploration Agency, SAR = synthetic aperture radar.

Source: JAXA, 2023.

Each feature contributes to rice classification in different ways. The Normalized Difference Vegetation Index (NDVI), for example, measures vegetation greenness by comparing the near-infrared and red bands of Sentinel-2 imagery. During the growth stage, NDVI values increase as the rice plants grow in size and density. The Normalized Difference Water Index (NDWI), which compares the green and near-infrared bands, highlights water features; this is particularly useful because rice fields are typically flooded during part of the season, making water-covered areas easy to detect.[26] The Normalized Difference Snow Index (NDSI) is designed to detect snow by comparing the green and shortwave infrared bands.[27] Although this may seem counterintuitive, NDSI also aids in identifying water areas and helps mask bright surfaces such as clouds and urban areas.

SAR imagery, known for its ability to penetrate clouds, is especially valuable in Southeast Asia where persistent cloud cover can limit coverage of optical imagery. The HH band in SAR imagery is sensitive to surface scattering, effectively detecting the roughness of flooded fields and distinguishing open water from other smooth surfaces. In contrast, the HV band is more responsive to volume scattering, capturing details about the internal structure of the rice plants. This allows for effective monitoring of the rice canopy throughout the growing season as it increases in volume.

26 S. K. McFeeters. 1996. The use of Normalized Difference Water Index (NDWI) in the delineation of open water features. *International Journal of Remote Sensing*. 17 (7). pp. 1425–1432.

27 W. Takeuchi, and Y. Yasuoka. 2004. Development of Normalized Vegetation, Soil and Water Indices Derived from Satellite Remote Sensing Data. *Journal of the Japan Society of Photogrammetry and Remote Sensing*. 43 (6). pp. 7–19.

The INAHOR Random Forest classifier further specifies 30 decision trees, each constructed using a bootstrap sample of the training data. During tree construction, a random subset of features, extracted from both the optical (NDVI, NDWI, NDSI) and radar (HH, HV backscatter coefficients) datasets is used at each node to determine the best split. This randomness reduces overfitting and increases the robustness of the classifier, with trees grown until a stopping criterion such as a minimum node size or maximum tree depth is met to ensure that the model captures complex patterns.

Figure 6 demonstrates the discriminative power of spectral indices, using NDVI as an example, in classifying rice. During peak growing periods (months March–April, July–August, and November–December), rice areas demonstrate higher NDVI values with minimal overlap to non-rice areas, allowing visual differentiation between classes. Conversely, in non-peak months (January–February, May–June, and September–October), greater visual overlap arises due to fallow fields or bare soil.

Figure 6: Distribution of Normalized Difference Vegetation Index Values Surrounding Rice and Non-Rice Observation Points by Months, 2024

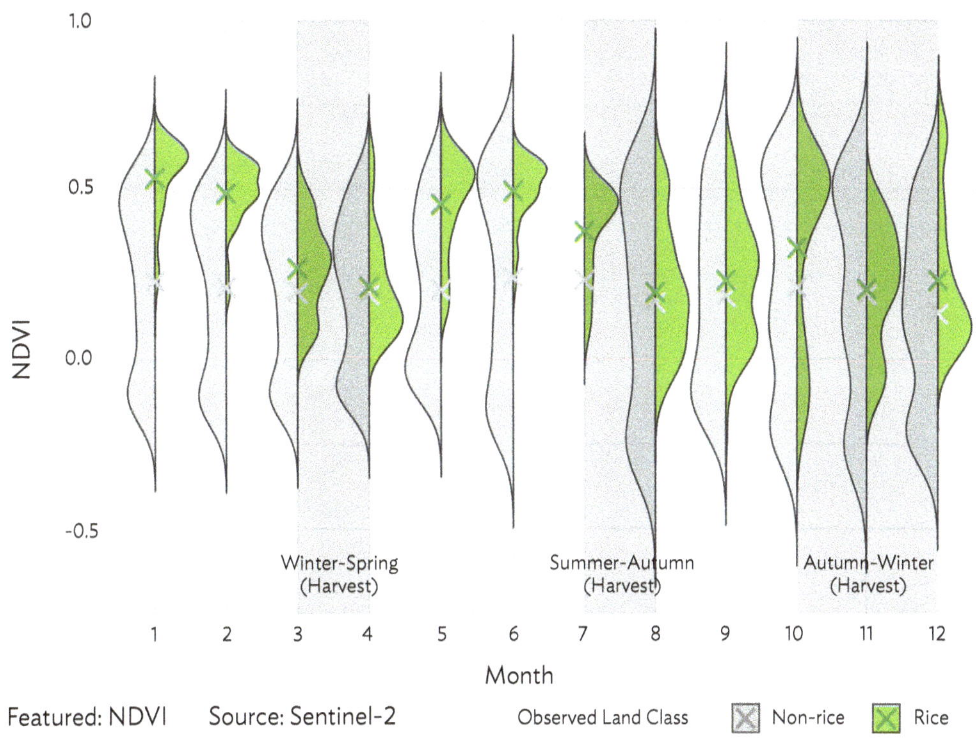

NDVI = Normalized Difference Vegetation Index.

Note: Violin plots depict the monthly distribution of NDVI values for field-observed rice (green) and non-rice (gray) locations, scaled by the number of observations. All ground-truth points were collected during August 2024; the figure therefore assumes that class labels (rice vs. non-rice) remain constant throughout the year.

Source: Asian Development Bank estimates. 2024.

We further assess the discriminative power of each feature considered by INAHOR to identify those that contribute most to the classification (Tables 3 and 4). Specifically, we examine the distributions of spectral values for NDVI, NDWI, NDSI, HH, and HV in both rice and non-rice areas, using the Wilcoxon statistical test to assess the significance of their differences.

For spectral indices derived from optical imagery, we identify seasonal patterns when the differences between rice and non-rice areas are most pronounced. Additionally, the table underscores a key limitation of optical imagery, where cloud cover can reduce the number of usable data points needed to reliably distinguish between the classes.

In SAR-derived indices, Table 4 shows that both the HH and HV backscatter coefficients consistently and significantly differ between rice and non-rice areas across all months, indicating strong discriminative power throughout the year.

Table 3: Median and Interquartile Range for Rice and Non-Rice by Month in 2024 (Optical Indices) with Wilcoxon Test *p*-Values

Month	NDVI			NDWI			NDSI		
	Rice	Non-Rice	p-value	Rice	Non-Rice	p-value	Rice	Non-Rice	p-value
1	0.58±0.13	0.27±0.51	0.000***	-0.51±0.11	-0.26±0.51	0.000***	-0.29±0.09	-0.21±0.36	0.000***
2	0.50±0.11	0.27±0.48	0.000***	-0.45±0.09	-0.26±0.51	0.000***	-0.23±0.06	-0.21±0.32	0.003**
3	0.28±0.20	0.24±0.43	0.059	-0.29±0.14	-0.25±0.43	0.048*	-0.24±0.07	-0.22±0.34	0.026*
4	0.17±0.20	0.23±0.41	0.722	-0.17±0.17	-0.23±0.46	0.965	-0.17±0.15	-0.22±0.34	0.123
5	0.50±0.17	0.21±0.42	0.000***	-0.45±0.18	-0.20±0.45	0.000***	-0.25±0.08	-0.20±0.33	0.000***
6	0.52±0.10	0.28±0.58	0.000***	-0.47±0.10	-0.27±0.53	0.000***	-0.26±0.06	-0.22±0.26	0.000***
7	0.45±0.16	0.23±0.00	N/A	-0.40±0.17	-0.24±0.00	N/A	-0.22±0.11	-0.17±0.00	N/A
8	0.17±0.28	0.22±0.63	0.819	-0.20±0.26	-0.22±0.52	0.544	-0.15±0.24	-0.17±0.41	0.159
9	0.25±0.28	0.17±0.50	0.272	-0.19±0.23	-0.16±0.36	0.419	-0.08±0.16	-0.14±0.43	0.294
10	0.48±0.50	0.22±0.62	0.007**	-0.42±0.39	-0.22±0.51	0.017*	-0.19±0.34	-0.17±0.40	0.920
11	0.24±0.29	0.21±0.54	0.641	-0.25±0.26	-0.21±0.49	0.823	-0.13±0.24	-0.18±0.33	0.233
12	0.15±0.25	0.13±0.52	0.076	-0.15±0.19	-0.13±0.53	0.056	-0.12±0.23	-0.20±0.38	0.923

NDSI = Normalized Difference Snow Index, NDVI = Normalized Difference Vegetation Index, NDWI = Normalized Difference Water Index.

Note: ***$p < 0.001$, **$p < 0.01$, *$p < 0.05$, N/A: Insufficient data.

Source: Asian Development Bank estimates. 2024.

Table 4: Median and Interquartile Range for Rice and Non-Rice by Month in 2024 ALOS-2 (Advanced Land Observing Satellite-2) Indices with Wilcoxon Test *p*-Values

Month	ALOS-HH			ALOS-HV		
	Rice	Non-Rice	p-value	Rice	Non-Rice	p-value
1	2930.90±1417.80	5084.52±2524.24	0.000***	960.37±512.80	2690.95±1495.24	0.000***
2	2959.17±1092.24	5058.61±2351.47	0.000***	1008.42±524.22	2627.53±1752.62	0.000***
3	3170.20±1663.63	4996.07±2566.61	0.000***	1323.92±677.82	2752.84±1506.43	0.000***
4	2365.79±1145.44	4954.32±2520.14	0.000***	946.90±523.78	2724.70±1749.86	0.000***
5	2494.64±1380.21	5116.50±2408.23	0.000***	842.46±485.84	2764.80±1672.89	0.000***
6	3184.57±1065.48	4908.53±2181.95	0.000***	1021.91±509.89	2645.29±1374.47	0.000***
7	3736.55±1441.66	5871.54±2607.80	0.000***	1402.13±591.88	2838.65±1540.97	0.000***
8	2894.13±1696.41	5298.21±2225.96	0.000***	1089.50±788.55	2707.36±1636.92	0.000***
9	2686.28±1559.84	5452.37±3368.92	0.000***	949.81±751.59	2811.85±1939.62	0.000***
10	3285.44±1470.44	5523.34±2312.87	0.000***	1178.56±596.63	2812.74±1736.64	0.000***
11	3050.75±1853.42	5192.94±2888.52	0.000***	1301.99±746.35	2611.21±1664.98	0.000***
12	2380.20±1432.92	5366.68±2719.79	0.000***	933.61±615.09	2860.98±1503.90	0.000***

ALOS-2-HH: HH polarization backscatter coefficient.
ALOS-2-HV: HV polarization backscatter coefficient.

Notes: ***$p < 0.001$, **$p < 0.01$, *$p < 0.05$.

Source: Asian Development Bank estimates. 2024.

For the classification process, each pixel within the composite image is evaluated by passing its corresponding feature vector through all 30 trees in the ensemble. Each tree independently assigns a label, rice (1) or non-rice (0), based on its learned decision rules. The final classification for each pixel is determined by majority voting across all trees, where the class receiving the most votes is assigned to the pixel.

The model's performance is validated using a fivefold cross-validation procedure. In this process, the training dataset is randomly partitioned into five subsets, and the model is iteratively trained on four subsets while testing on the remaining one. Post-classification, the output is further refined with a majority filtering and masking using ancillary datasets, e.g., WorldCover and MODIS.

Measuring Uncertainty of INAHOR Estimates

Since INAHOR's Random Forest outputs lack inherent measures of uncertainty, this study implements a jackknife bootstrapping procedure to calculate sampling error specifically for estimates produced by INAHOR. This enables a direct comparison between total rice area estimates from INAHOR and those derived using expansion and ratio estimators from the area sampling frame method.

The jackknife bootstrapping procedure systematically removes one grid at a time from the dataset, resulting in 34 jackknife replicates (one for each grid in the sample). For example, Replicate 1 is created by excluding GRID 1 (using grids 2, 3, 4, ..., 34), Replicate 2 by excluding GRID 2 (using grids 1, 3, 4, ..., 34), and so on, until Replicate 34, which omits GRID 34 (using grids 1, 2, 3, ..., 33). Each replicate is then processed through the INAHOR platform to generate an estimate of the total area planted. This systematic approach allows us to quantify the variability in the estimates and provides a robust measure of sampling errors.

The jackknife estimates and its variance, which provides a measure of uncertainty, is calculated as:

$$\hat{Y}_{JK} = \frac{1}{34} \sum_{i=1}^{34} \hat{Y}_{(i)} \qquad (13)$$

$$s^2\left(\hat{Y}_{JK}\right) = \frac{1}{34(33)} \sum_{i=1}^{34} \left(\hat{Y}_{(i)} - \hat{Y}_{JK}\right)^2 \qquad (14)$$

Results

The following section presents the results of rice area estimates derived from INAHOR and three model-assisted estimators: the Expansion Estimator, the Ratio Estimator, and the Ratio Estimator Excluding Segments with Matching Issues. These estimators are based on the observed percentage of rice within 50 m × 50 m segments collected during the ground truth survey. The INAHOR and model-assisted estimator results are finally compared against NSO administrative data for external validation and assessment. A summary of results is provided in Table 5.[28]

Table 5: Summary of Area Estimates Across Methods

Estimator	Estimate (km^2)	CV (%)	Comparison with NSO (%)
Remote Sensing (INAHOR)	2,500.1	0.19	+9.4
Expansion Estimator	2,517.6	8.04	+10.2
Ratio Estimator	2,444.5	5.02	+7.0
Adjusted Ratio Estimator	2,501.1	0.30	+9.5
NSO Administrative Report	2,284.8	-	-

CV = Coefficient of Variation, INAHOR = International Asian Harvest mOnitoring system for Rice, km^2 = square kilometer, NSO = National Statistics Office.

Source: Asian Development Bank compilation of estimates from multiple analyses. 2024.

28 District-level estimates are presented in Appendix 1.

The sequence of estimators presented in Table 4 progressively improves the precision of the rice area estimates by incorporating auxiliary data from the JAXA LULC, used initially for the construction of the sample frame. The expansion estimator directly applies observed rice proportions to the entire sample frame. The ratio estimator introduces auxiliary information from the JAXA land cover map, using the correlation between remotely sensed and observed rice proportions to lower segment-to-segment variability, and thus reducing the CV. The adjusted ratio estimator further improves precision by excluding segments with spatial classification inconsistencies, yielding the lowest CV. We observe the adjusted ratio as the default estimate based on the relative precision of the three estimators.

Figure 7 presents the rice area map for An Giang Province, generated by the INAHOR platform using ground truth data collected in August 2024. Of the province's 3,532.59 km^2, 2,511.62 km^2 (71.1%) are classified as rice, while 1,020.97 km^2 (28.9%) are classified as non-rice. The map, produced at a resolution of 50 meters per pixel, estimates area by counting the number of pixels designated as rice.

Accuracy metrics, averaged over fivefold cross-validation, indicate robust performance: Overall accuracy is 95.3%, with producer accuracies of 93.24% for rice and 97.18% for non-rice, and user accuracies of 96.94% for rice and 93.79% for non-rice. The kappa coefficient of 0.91 demonstrates high agreement between the classification and ground truth data.

Figure 7: Rice Cultivation Map of An Giang Province, Summer–Autumn 2024

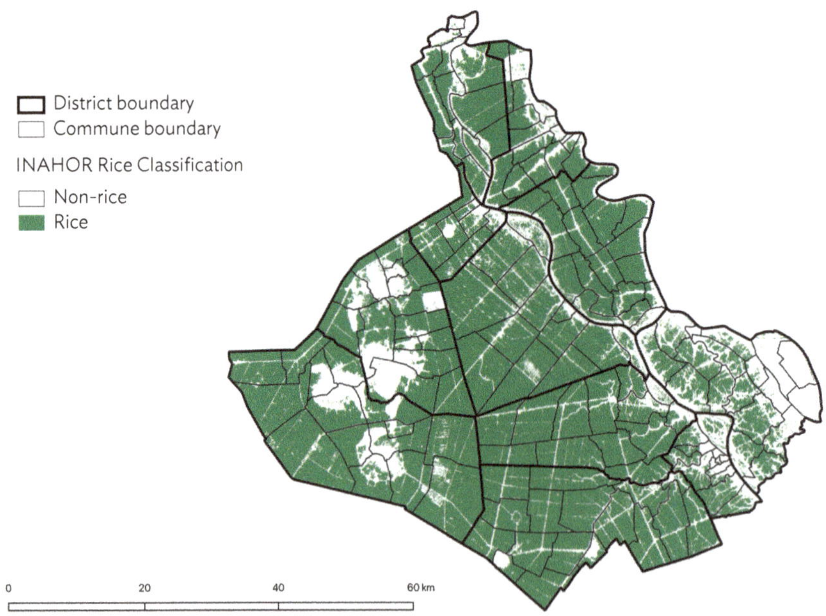

INAHOR = International Asian Harvest mOnitoring system for Rice, km^2 = square kilometer.

Note: The map reflects the administrative boundaries of An Giang Province prior to its recent merger in June 2025. All classification and analysis are based on the former provincial configuration.

Source: Asian Development Bank visualization using QGIS 3.34.14. 2024. Manila.

Using the jackknife bootstrapping procedure, we derive the uncertainty metrics for rice area estimation using INAHOR. The estimated rice area is 2,500.10, with a standard error of 4.72, a CV of 0.19%, and a 95% confidence interval ranging from 2,490.50 to 2,509.70. The low CV of 0.19% indicates a high degree of precision and consistency in the classification across replicates.

Utilizing these uncertainty metrics, we map the average and variance of each replicate to highlight areas within the study area where the classification procedure may have struggled to classify rice areas consistently. Figure 8 depicts two mapping outputs related to these uncertainty metrics. The map on the left provides the mean of the binary (rice [1] and non-rice [0]) classification, where a value closer to 1 implies a classification that is performing consistently across replicates. The map on the right provides the variance estimate across the replicates. Areas with higher variance depict areas where the classification may have not performed consistently. Areas with higher variance tend to be representative of areas with mixed pixels (bordering between two land classes) or areas with a higher mix of land use classes. Cho Moi district expressed a relatively higher degree of variance, as field observations noted the presence of mixed cropping or where rice fields have been converted to other uses. Noting areas of high variance helps not only to refine the classification but also to inform future ground truth survey exercises.

Figure 8: Uncertainty Map, An Giang Province, Summer–Autumn 2024

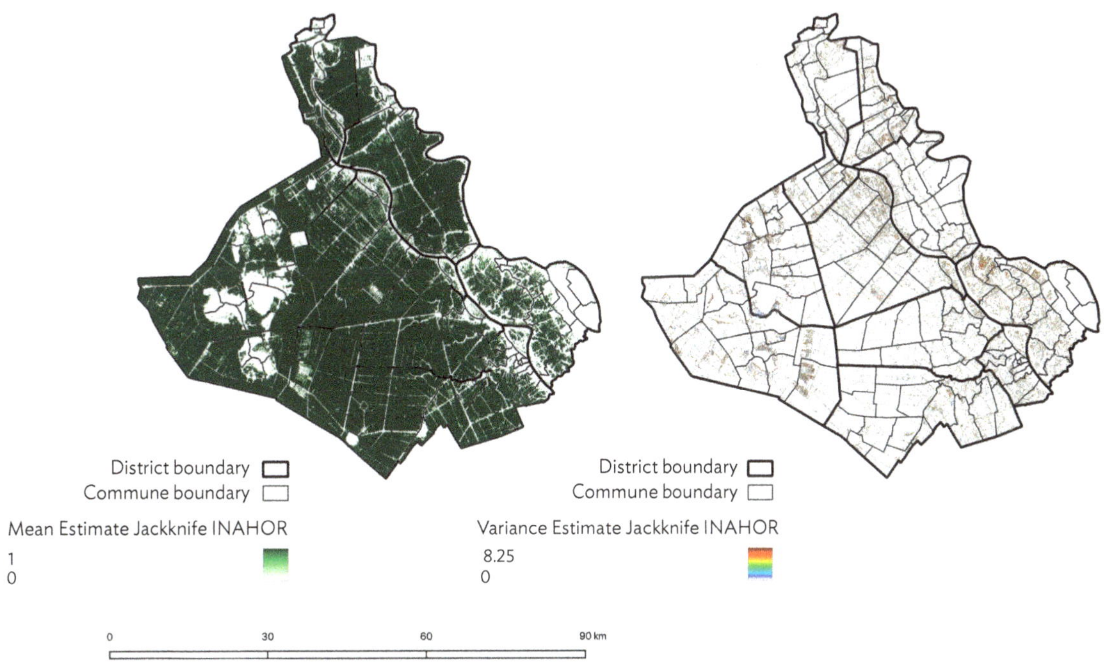

INAHOR = International Asian Harvest mOnitoring system for Rice.

Note: The map reflects the administrative boundaries of An Giang Province prior to its recent merger in June 2025. All classification and analysis are based on the former provincial configuration.

Source: Asian Development Bank visualization using QGIS 3.34.14. 2024. Manila.

Figure 9 presents rice area estimates derived from different rice area estimation methods, along with their respective CVs and percentage differences from administrative reports. The Expansion estimator produces a rice area estimate of 2,517.63 km^2, with a CV of 8.04% and a 10.19% difference from NSO administrative reports. The CV falls within the target level of precision, demonstrating the effectiveness of using area sampling frames derived from remotely sensed maps for sample size determination within a given sampling design.

Figure 9: Comparison of Rice Cultivation Areas by Method

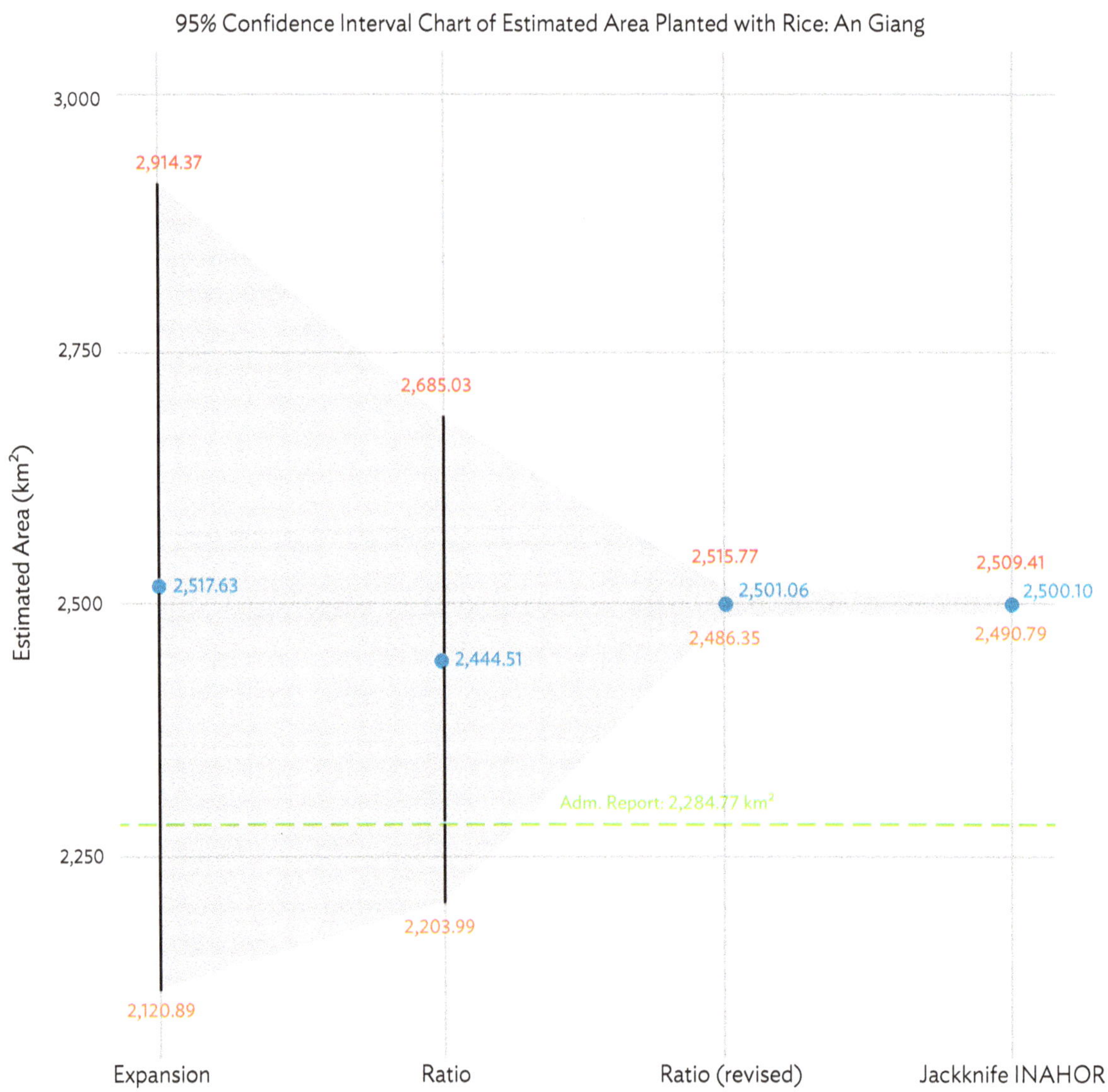

INAHOR = International Asian Harvest mOnitoring system for Rice, km^2 = square kilometer.
Source: Asian Development Bank estimates. 2024.

The ratio estimator provides a slightly lower estimate of 2,444.51 km^2 compared to the Expansion estimator, with improved precision (CV = 5.02%) and a 6.99% difference from NSO administrative records. However, some discrepancies were observed in the matching of segment-level characteristics, particularly in the total area planted with rice. These mismatches may introduce inconsistencies that could impact the precision of the ratio estimates, leading to larger-than-expected differences in some cases.

The adjusted ratio estimator further refines the estimate to 2,501.06 km^2, resulting in the lowest CV of 0.3% and a 9.47% difference from administrative reports. The adjusted ratio estimator yields the highest precision increase among design-based estimators, surpassing both the expansion estimator (CV = 8.04%) and the original ratio estimator (CV = 5.02%). This reduction suggests that the exclusion of problematic segments has effectively reduced uncertainty.

In summary, all estimation methods, including expansion, ratio, and adjusted ratio estimators, yield rice area estimates that exceed the NSO administrative records, suggesting a potential underestimation in official figures that may require further validation. At the province level, the confidence intervals for the expansion and ratio estimators overlap with the administrative data, indicating that the differences between these estimates are not statistically significant at the 5% level. Among the four estimators, the adjusted ratio estimator and the jackknife INAHOR estimator produce closely aligned estimates, converging around 2,500 km^2 of rice area.

Discussion

The high classification accuracy of the rice area maps and the precision of the sample-based estimates highlight the strength of combining design-based sampling with remote sensing for rice area estimation in An Giang Province. These findings indicate the potential of this methodology for producing reliable and scalable crop area statistics. However, they also reveal some operational and methodological challenges that require further consideration in scaling the methodology.

Area survey and remote sensing methods proved highly effective for estimating rice areas in An Giang Province and its districts. By applying a design-based, objective approach that integrated remote sensing data, the study produced reliable provincial-level estimates with a CV below 10%. The sample size was sufficient to support both the estimation of provincial results, and reliable district-level results with the use of model-assisted estimators.

Design-based, finite population sampling principles can be effectively applied to ground truth surveys, moving beyond traditional "rule of thumb". Moving beyond rule-of-thumb methods, we establish more robust approaches for determining sample size and allocating sampling points within the study area. This method demonstrates that high-accuracy mapping products can be achieved through sound sampling principles, while also improving the efficiency of ground truth data collection and thereby reducing operational costs.

LULC maps are valuable data sources for constructing area sampling frames and enhancing model-assisted estimation. Just as high-quality census data strengthen sample surveys, this study demonstrates how LULC maps can support the design of area frames by enabling variance estimation and informed sample size determination. Additionally, LULC data were used as auxiliary variables in a model-assisted estimator to improve precision in area estimates.

High-quality, timely ground truth data are critical for producing accurate crop maps. Classification models are highly sensitive to the accuracy of labels and the alignment of reference time periods; deviations in either can reduce mapping quality. This challenge was evident during our field exercise, where enumerators occasionally struggled with labeling due to rice being at varying growth stages. To address this, future efforts may consider collecting ground truth data consistently throughout the growing season rather than concentrating solely on the harvest period, thereby improving labeling accuracy and overall map reliability.

The availability of user-friendly, cost-efficient crop monitoring systems helps countries sustainably use satellite imagery and remote sensing technologies. Building sustainability in the use of remote sensing involves reducing barriers to adoption, transforming research tools into streamlined products with user-friendly interfaces and lower computational demands. Advocacy is also essential to raise awareness about the data inputs these tools require, and how official data producers can collect such data, either through stand-alone surveys or by integrating them into existing statistical products. The latter approach can enhance efficiency and resource use.

Discrepancies between rice area estimates highlight potential underreporting in administrative data. Estimates derived from sampling and remote sensing consistently exceeded those reported through administrative data, suggesting systematic underreporting. Further validation may be required to assess the reliability and potential biases of administrative data with respect to remote sensing estimates. The discrepancy likely stems from differences in reference dates and minimum threshold criteria used in administrative reporting, including variations in reporting periods and the exclusion of areas that fall below the administrative reporting threshold.

Future research could apply this methodology to other regions to evaluate its robustness and develop a reproducible framework for the Asia and Pacific region. Given that field data collection represents the most significant cost component, future efforts may also explore cost-effective technological alternatives, such as high-resolution imagery from other satellites or unmanned aerial vehicles, to acquire imagery sufficient for labeling by subject matter experts.

While remote sensing and area frame methods offer improved timeliness and accuracy of crop area estimation, they should be viewed as complements to—rather than a substitute for—existing administrative reporting systems. Objective measurements from remote sensing estimates additionally offer valuable auxiliary data for improving the precision of agricultural statistics through model-assisted estimation, variance reduction, and enhanced small-area estimation. Administrative data remain essential for capturing farmer-level socioeconomic details such as input use and management practices, which cannot be directly observed from satellite imagery.

Remote sensing offers substantial efficiencies in both time and resource allocation, and is particularly beneficial for rapid, time-sensitive assessments. In this study, ground truth data collection and processing for remote sensing estimates required approximately 1 week, involving only a small team of enumerators. By contrast, administrative methods typically rely on monthly field observations conducted by a network of local district staff, which often take several weeks to compile. Moreover, the use of free and open-source tools such as GEE and QGIS concentrates costs primarily on labor and capacity building rather than expensive software or infrastructure.

To effectively integrate remote sensing-based rice mapping into the national agricultural statistics system, ground truth surveys can be institutionalized either as stand-alone statistical activities or embedded within existing agricultural data collection efforts. These may include crop-cutting experiments or production surveys, which offer opportunities for greater cost-efficiency. This integration requires strengthening capacity and coordination between GIS and agricultural statistics units to ensure consistency in sampling frames, alignment with cropping seasons, harmonized questionnaire design, and proficiency with free and open-source tools such as QGIS and GEE.

Conclusion

The mapping of rice cultivation areas using remote sensing methods benefits from the application of a robust sampling methodology for ground truth data collection. Access to high-quality ground truth data is essential for producing accurate crop area maps, and the integration of robust sampling approaches ensures more efficient and cost-effective sample sizes and allocation.

The methodology also leverages various geospatial data products, such as the JAXA LULC map, to support sampling design, demonstrating how traditional sampling frameworks can be enhanced through satellite-derived data. Specifically, we illustrate the use of LULC maps to construct the area frame, inform sample size and allocation, and improve the precision of sample-based rice area estimates.

User-friendly tools such as INAHOR are also pivotal for the adoption of these methodologies as they help to bridge the skill gap between official statistics and remote sensing. Accessibility of these systems also ensures their sustainability.

The precision of our estimates and their discrepancies with administrative reporting sources further highlights the methodology's value in validating existing data. By incorporating alternative data sources, we can better identify localities where discrepancies occur and improve the quality of crop area statistics.

Lessons from this study will support the scaling of the methodology and benefit future applications through further testing across different geographies and cropping seasons. Overall, more frequent monitoring and the continued use of remote sensing tools can help governments become more data-driven and deliver timely information to support policy needs, especially in a world increasingly challenged by food security, climate change, and environmental events.

APPENDIX 1

Tables of Rice Area Estimates by Province and District

Table A1.1. Expansion Estimator of Total Area Planted with Rice (km^2)

Province/District	Estimate	SE	CV(%)	LL	UL
An Giang	2,501.06	7.46	0.30	2,486.44	2,515.68
Thành phố Long Xuyên (883)	55.28	33.87	61.27	-15.37	125.92
Thành phố Châu Đốc (884)	69.42	1.42	2.04	66.46	72.38
Huyện An Phú (886)	228.25	1.78	0.78	224.52	231.97
Thị xã Tân Châu (887)	71.64	27.73	38.71	13.80	129.49
Huyện Phú Tân (888)	158.21	157.71	99.69	-170.77	487.18
Huyện Châu Phú (889)	407.13	16.38	4.02	372.96	441.31
Huyện Tịnh Biên (890)	304.55	2.40	0.79	299.54	309.56
Huyện Tri Tôn (891)	460.38	49.94	10.85	356.20	564.55
Huyện Châu Thành (892)	235.41	103.18	43.83	20.18	450.64
Huyện Chợ Mới (893)	63.75	21.19	33.24	19.55	107.95
Huyện Thoại Sơn (894)	463.63	17.38	3.75	427.37	499.90

CV = Coefficient of Variation, km^2 = square kilometer, LL = Lower Limit, SE = Standard Error, UL = Upper Limit.

Table A1.2. Ratio Estimator of Total Area Planted with Rice (km^2)—All Sample Observations

Province/District	Estimate	SE	CV(%)	LL	UL
An Giang	2,444.51	122.77	5.02	2,203.87	2,685.14
Thành phố Long Xuyên (883)	55.33	2.72	4.91	49.67	62.12
Thành phố Châu Đốc (884)	50.75	3.63	7.16	43.16	40.60
Huyện An Phú (886)	156.42	9.01	5.76	137.62	178.33
Thị xã Tân Châu (887)	76.16	9.41	12.36	56.52	66.51
Huyện Phú Tân (888)	224.02	36.88	16.46	147.09	274.72
Huyện Châu Phú (889)	347.49	18.22	5.24	309.47	378.65
Huyện Tịnh Biên (890)	219.29	0.08	0.03	219.13	209.78
Huyện Tri Tôn (891)	429.50	19.37	4.51	389.09	430.37
Huyện Châu Thành (892)	288.29	3.16	1.09	281.70	280.91
Huyện Chợ Mới (893)	203.20	111.76	55.00	-29.93	591.00
Huyện Thoại Sơn (894)	394.07	17.75	4.50	357.05	411.82

CV = Coefficient of Variation, km^2 = square kilometer, LL = Lower Limit, SE = Standard Error, UL = Upper Limit.

Table A1.3. Adjusted Ratio Estimator of Total Area Planted with Rice (km^2)—Excluding 25 Observations with Large Matching Problems

Province/District	Estimate	SE	CV(%)	LL	UL
An Giang	2,501.06	7.46	0.30	2,486.44	2,515.68
Thành phố Long Xuyên (883)	55.33	2.72	4.91	49.67	62.12
Thành phố Châu Đốc (884)	74.14	1.11	1.50	71.83	77.75
Huyện An Phú (886)	154.24	1.66	1.08	150.78	158.27
Thị xã Tân Châu (887)	108.58	0.73	0.67	107.07	109.00
Huyện Phú Tân (888)	244.18	0.00	0.00	244.18	242.96
Huyện Châu Phú (889)	340.89	1.94	0.57	336.85	332.39
Huyện Tịnh Biên (890)	219.29	0.08	0.03	219.13	209.78
Huyện Tri Tôn (891)	464.69	0.31	0.07	464.04	461.10
Huyện Châu Thành (892)	288.29	3.16	1.09	281.70	280.91
Huyện Chợ Mới (893)	144.25	0.00	0.00	144.25	138.71
Huyện Thoại Sơn (894)	407.16	5.47	1.34	395.76	413.14

CV = Coefficient of Variation, km^2 = square kilometer, LL = Lower Limit, SE = Standard Error, UL = Upper Limit.

Table A1.4. Jackknife Estimator from Estimated Area Using International Asian Harvest mOnitoring system for Rice (INAHOR)

Province/District	Estimate	SE	CV(%)	LL	UL
An Giang	2,500.10	4.72	0.19	2,490.85	2,509.35
Thành phố Long Xuyên (883)	50.65	0.25	0.49	50.16	51.13
Thành phố Châu Đốc (884)	73.86	0.24	0.32	73.39	74.33
Huyện An Phú (886)	151.00	0.26	0.17	150.48	151.51
Thị xã Tân Châu (887)	105.51	0.35	0.33	104.83	106.20
Huyện Phú Tân (888)	241.80	0.37	0.15	241.08	242.51
Huyện Châu Phú (889)	355.95	0.82	0.23	354.34	357.57
Huyện Tịnh Biên (890)	225.19	0.59	0.26	224.03	226.34
Huyện Tri Tôn (891)	476.87	0.76	0.16	475.37	478.37
Huyện Châu Thành (892)	293.30	0.34	0.11	292.64	293.96
Huyện Chợ Mới (893)	127.27	1.02	0.80	125.28	129.27
Huyện Thoại Sơn (894)	398.71	0.55	0.14	397.64	399.78

CV = Coefficient of Variation, km^2 = square kilometer, LL = Lower Limit, SE = Standard Error, UL = Upper Limit.

APPENDIX 2

Ground Truth Survey Questionnaire Form

Japan Fund for Prosperous and Resilient Asia and the Pacific

From the People of Japan

CONFIDENTIAL

2024 RICE AREA AND YIELD SURVEY (PILOT AN GIANG PROVINCE)
FORM 1 – GROUND TRUTH OBSERVATION OF SAMPLE GRID

NATIONAL STATISTICS OFFICE, VIET NAM

The information being collected under the Statistics Act 2003 and will be kept strictly confidential and will be used for statistical purposes only.

START: *Travel to the your sample point (A03) and note the boundary (A04-7) of the assigned observation area.*

SECTION A. COVER (pre-filled by NSO):

A01. GRID ID (1 km x 1 km)		
A02. SAMPLE ID		
A03. SAMPLE POINT (CENTER)	N	
	E	
A04. SAMPLE GRID BOUNDARY – UPPER LEFT	N	
	E	
A05. SAMPLE GRID BOUNDARY – UPPER RIGHT	N	
	E	
A06. SAMPLE GRID BOUNDARY – LOWER LEFT	N	
	E	
A07. SAMPLE GRID BOUNDARY – LOWER RIGHT	N	
	E	

SIGNATURE			
	Interviewer	Supervisor	Headquarters
Signature			
Full Name			
Phone number			

SECTION B: OBSERVATION OF SAMPLE GRID

B01. Are you able to navigate to the sample point?	◯ 0. NO ◯ 1. YES à Skip to B02									
B01a. Why were you not able to reach the sample point?	TEXT									
B02. GPS coordinate of observation point	E									
	N									
B03. Date and time of observation	DD/MM/YYYY HH:MM:SS									
B04. Major observed land use in sample grid	◯ 1. Rice → Skip to B05 ◯ 2. Other crops ◯ 3. Urban → Skip to B05 ◯ 4. Water → Skip to B05									
B04a. If no rice is planted, is most of the observed land used for:	◯ 1. Annual crops ◯ 2. Perennial crops ◯ 3. Livestock grazing ◯ 4. Forestry ◯ 5. Aquaculture ◯ 6. Other, specify									
B05. Standing at the observation point take four landscape photos, pointing:										
B05a. NORTH	PHOTO									
B05b. EAST	PHOTO									
B05c. SOUTH	PHOTO									
B05d. WEST	PHOTO									
B06. Percent of observed area covered with Rice (0-100%)										
B07. Other observations	TEXT									

SECTION C: PARCEL AREA MEASUREMENT

C01. Number of land parcels overlapping sample grid			
PARCEL ROSTER			
C02. Parcel ID	**C03.** Sketch of parcel and dimensions	**C04.** Parcel Area (m^2)	**C05.** Land Use of Parcel ◯ 1. Rice ◯ 2. Other crops ◯ 3. Urban ◯ 4. Water
01			
02			
03			
04			

APPENDIX 3

International Asian Harvest mOnitoring system for Rice Technical Specification

This Appendix outlines the processes implemented by the International Asian Harvest mOnitoring system for Rice (INAHOR) system within Google Earth Engine (GEE) for rice area classification. The INAHOR system uses a Random Forest classifier to map rice paddy fields using a combination of ALOS-2 PALSAR-2 synthetic aperture radar (SAR) data and Sentinel-2 optical imagery within the GEE environment.

INAHOR processes are conducted through nine stages from data acquisition to the export of the final mapping product.

1. Data Acquisition

The INAHOR tool requires several key input files and specifications to perform its rice monitoring and mapping process effectively. The input files include three primary shapefiles. First, a shapefile defining the Area of Interest (AOI) is used to delineate the geographic extent for which the analysis will be conducted. Second, a shapefile containing the training data is required. This shapefile includes ground-truth information, such as land cover classes, which the Random Forest classifier utilizes to learn and differentiate between different types of land cover based on spectral characteristics and radar backscatter properties. Third, a shapefile representing the administrative boundary is used to aggregate classified areas by administrative units such as districts or communes.

Specification of International Asian Harvest mOnitoring system for Rice User Inputs

1. Area of Interest (AOI). A geospatial shapefile (.shp, .shx, .dbf) defining the study area's extent. In this study, the AOI corresponds to the administrative boundary of An Giang Province, Viet Nam. For this study, the shapefile was provided by National Statistics Office of Viet Nam in July 2024.

2. Training data. A geospatial vector shapefile (.shp, .shx, .dbf) containing training point locations for rice and non-rice classifications. Each point must include: a unique identifier, latitude, longitude, and a rice classification label, where 1 indicates rice and 0 indicates non-rice. Training data was collected from for this study in August 2024 per the sample design outlined in Section X.

3. Administrative Boundary. A geospatial vector shapefile (.shp, .shx, .dbf) representing administrative boundaries within the AOI. This file is used to aggregate classified rice and non-rice areas at the administrative level for final reporting.

4. Period for ALOS-2 data. The user must specify the start and end dates (YYYY-MM-DD format) for satellite imagery acquisition, corresponding to the rice-growing season under analysis. The study aligns the start and end date with the Summer-Autumn rice season from 1 April–31 August 2024.

continued

continued

5. Column name for land cover class. The attribute name in the training dataset that stores the rice classification label, where: 1 = Rice, 0 = Non-Rice.
6. Resolution of the classified image. The spatial resolution of the classification output is set to 50 meters, meaning each pixel represents a 50 m x 50 m area. 50 meters corresponds with the area observed during ground truth field data collection.
7. Resolution of the export image. The final exported image is resampled to a 25-meter resolution, producing a finer spatial resolution during export.
8. Administrative boundary name. The attribute field in the administrative boundary shapefile that contains the name of each administrative unit, used for area statistics reporting. In this study, rice area is reported at the province and district level

The INAHOR further retrieves satellite data from the GEE catalog for the specified AOI and start and end dates from two sources:

The first dataset, ALOS-2 PALSAR-2 Level 2.2 ScanSAR, provides SAR imagery from JAXA's ALOS-2 satellite. The second dataset, Sentinel-2 Multispectral Instrument, supplies high-resolution optical imagery from the European Space Agency (ESA)'s Sentinel-2 satellite.

2. Pre-Processing

In the Pre-Processing stage, spectral indices are calculated from ALOS-2 PALSAR-2 SAR data and Sentinel-2 (S2) optical imagery to prepare the datasets for classification.

For Sentinel-2, three spectral indices are computed to extract vegetation and water-related information. The Normalized Difference Vegetation Index (NDVI) is calculated from the near-infrared (B8) and red (B4) bands to highlight vegetation health. The Normalized Difference Water Index (NDWI) is derived from the green (B3) and shortwave infrared (B11) bands to identify water bodies (McFeeters, 1996). Similarly, the Normalized Difference Snow Index (NDSI) is generated from shortwave infrared (B11) and near-infrared (B8) bands based on Takeuchi et al. These indices are computed for each Sentinel-2 image, and their median values across the dataset are used to create a composite image containing NDVI, NDWI, and NDSI.

For ALOS-2 PALSAR-2 data, a series of corrections and transformations are applied. A no-data filter removes pixels where backscatter values are missing. The digital number-to-decibel conversion transforms raw digital number values into decibel units. A smoothing filter is then applied using a median focal operation to reduce noise while preserving edge information. The horizontal-horizontal (HH) and horizontal-vertical (HV) polarization bands are processed further by calculating their sum (Sum) and difference (Def) to enhance contrast between land cover types.

After pre-processing, maximum, minimum, and mean values are computed for the HH, HV, Sum, and Def bands, while the range (difference between maximum and minimum values) is calculated for HH and HV to capture temporal variation in backscatter. The final dataset integrates the ALOS-2 processed outputs with the Sentinel-2 indices, producing a composite image.

3. Training Data Creation

The training data creation stage involves a spatial join between the composite image from stage 2 and the training data shapefile inputted by the analyst in stage 1. The composite image is overlayed with the training points, which are labeled as rice (1) or non-rice (0). Each point gets the values from the corresponding pixel, adding both Sentinel-2 and ALOS-2 attributes to the training data point. This process is done at the classification image's spatial resolution (defined by the scale parameter) to keep the data consistent.

4. Specification of the Random Forest Classifier

INAHOR utilizes the Random Forest model to classify rice areas. The Random Forest model, implemented through the smileRandomForest function, comprises 30 decision trees. The classifier is trained using the training samples generated in the preceding step, wherein each point comprises spectral attributes, along with the corresponding rice (1) or non-rice (0) classification label.

5. Classification

In the Classification stage, the trained Random Forest classifier is applied to the processed composite dataset from stage 3, generating a land cover classification image where each pixel is labeled as either rice (1) or non-rice (0).

6. Post-Classification

In the post-classification stage, a masking process is applied using external datasets to improve accuracy and remove noise. First, the MODIS Water Mask (MODIS/MOD44W) is used to exclude permanent water bodies, masking non-arable regions. Next, the ESA WorldCover dataset is used to filter out non-agricultural land, retaining only cropland (40) and wetland (90) classifications for analysis.

Further refinement is conducted through a majority filter, which smooths the classification output by replacing isolated misclassified pixels with the dominant neighboring class. This filtering process, applied with a 5x5 pixel kernel, helps reduce speckling and enhances spatial consistency. Finally, the classified image is clipped to the AOI, ensuring that only results within An Giang Province are included.

7. Accuracy Assessment

To evaluate the performance and generalization capability of the classifier, we implemented a fivefold cross-validation procedure. The training data were randomly partitioned into training and testing subsets with a 50-50 split, ensuring that both subsets were representative of the overall data distribution. This random partitioning was repeated five times, each with a different random seed, to mitigate the effects of any data split and to provide a more robust estimate of classifier performance. For each iteration, the classifier was retrained, and predictions were made on both the training and testing subsets. Confusion matrices were generated for each iteration to compute metrics such as overall accuracy, which provided insights into both the classifier's fitting to the training data and its predictive power on unseen data.

References

Belgiu, M., and O. Csillik. 2018. Sentinel-2 Cropland Mapping Using Pixel-Based and Object-Based Time-Weighted Dynamic Time Warping Analysis. *Remote Sensing of Environment*. 204. pp. 509–523.

Burgard, A. et al. 2024. *Plotting from Above: Enhancing Agricultural Mapping in Asia and the Pacific*. https://www.adb.org/sites/default/files/publication/976906/plotting-above-agricultural-mapping-asia-pacific.pdf.

Carletto, C. et al. 2016. *Cheaper, Faster, and More than Good Enough: Is GPS the New Gold Standard in Land Area Measurement ?*. World Bank. https://documents.worldbank.org/en/publication/documents-reports/documentdetail/257481469116364362/Cheaper-faster-and-more-than-good-enough-is-GPS-the-new-gold-standard-in-land-area-measurement.

Congalton, R. G. 1988a. A Comparison of Sampling Schemes Used in Generating Error Matrices for Assessing the Accuracy of Maps Generated from Remotely Sensed Data. *Photogrammetric Engineering and Remote Sensing*. 54 (5). pp. 593–600.

Congalton, R. G. 1988b. Using Spatial Autocorrelation Analysis to Explore the Errors in Maps Generated from Remotely Sensed Data. *Photogrammetric Engineering and Remote Sensing*. 54 (5). pp. 587–592.

Congalton, R. G. 1991. A Review of Assessing the Accuracy of Classification of Remotely Sensed Data. *Remote Sensing of Environment*. 37 (1). pp 35–46.

Defourny, P., et al. 2019. Near Real-Time Agriculture Monitoring at National Scale at Parcel Resolution: Performance Assessment of the Sen2-Agri Automated System in Various Cropping Systems around the World. *Remote Sensing of Environment*. 221. pp. 551–568.

de Simone, L., W. Ouellette, P. Gennari. (n.d.). *Operational Use of EO Data for National Land Cover Official Statistics in Lesotho*.
https://www.mdpi.com/2072-4292/14/14/3294.

Durante, A.C. et al. 2018. Improving Paddy Rice Statistics Using Area Sampling Frame Technique. *ADB Economics Working Paper Series*. No. 565. Asian Development Bank.

Earth Engine Data Catalog. (n.d.). Google for Developers. https://developers.google.com/earth-engine/datasets (accessed 11 June 2025).

European Space Agency (ESA). 2023. *Sentinels for Agricultural Statistics – Final Report*. https://eo4society.esa.int/wp-content/uploads/2023/02/esa-sentinels-for-agricultural-statistics.pdf.

Fitzpatrick-Lins, K. 1981. Comparison of Sampling Procedures and Data Analysis for a Land-Use and Land-Cover Map. *Photogrammetric Engineering and Remote Sensing*. 47 (3). pp. 343–351.

Food and Agriculture Organization of the United Nations (FAO). 2018. *EOSTAT: Trusted Methods for the Use of EO Data for Crop Type Mapping and Crop Yield Forecasting*. https://openknowledge.fao.org/bitstreams/32767090-b0be-40a2-9831-c1928fd13a6d/download.

Gallego, F. J. 2004. Remote Sensing and Land Cover Area Estimation. *International Journal of Remote Sensing*. 25 (15). pp. 3019–3047.

Gandharum, L. et al. 2021. Remote Sensing Versus the Area Sampling Frame Method in Paddy Rice Acreage Estimation in Indramayu Regency, West Java Province, Indonesia. *International Journal of Remote Sensing*. . pp. 1738–1767.

Geo-Informatics and Space Technology Development Agency (GISTDA). (2025). https://www.gistda.or.th/home.php (accessed 10 June 2025).

Japan Aerospace Exploration Agency (JAXA). 2023. *High-Resolution Land-Use and Land-Cover Map of Vietnam for 2020 (Version 23.09)*. https://www.eorc.jaxa.jp/ALOS/en/dataset/lulc/lulc_vnm_v2309_e.htm.

McFeeters, S. K. 1996. The Use of the Normalized Difference Water Index (NDWI) in the Delineation of Open Water Features. *International Journal of Remote Sensing*. 17 (7). 1425–1432.

Ministry of Agriculture and Rural Development (MARD). 2024.

Moraes, D. et al. 2021. Influence of Sample Size in Land Cover Classification Accuracy Using Random Forest and Sentinel-2 Data in Portugal. *2021 IEEE International Geoscience and Remote Sensing Symposium IGARSS*. Brussels: IEEE.

National Statistics Office of Viet Nam (NSO). 2024. *Statistical Yearbook of 2023*. https://www.nso.gov.vn/en/default/2024/07/statistical-yearbook-of-2023/.

Nusser, S. and E. Klaas. 2003. *Survey Methods for Assessing Land Cover Map Accuracy*. U.S. Geological Survey. https://pubs.usgs.gov/publication/70025183.

Philippine Rice Information System (PRiSM) 2025. Data Products. https://prism.philrice.gov.ph/ (accessed 11 June 2025).

Remote Sensing Technology Center of Japan (RESTEC). 2024. *Use Case: Agriculture*. https://www.restec.or.jp/en/service/agriculture/index.html.

Rotairo, L. et al. 2019. *Use of Remote Sensing to Estimate Paddy Area and Production: A Handbook*. Manila: Asian Development Bank. https://doi.org/10.22617/TIM190098-3.

Shetty, S. et al. 2021. Assessing the Effect of Training Sampling Design on the Performance of Machine Learning Classifiers for Land Cover Mapping Using Multi-Temporal Remote Sensing Data and Google Earth Engine. *Remote Sensing*. 13(8). 1433.

Space Applications for Environment. (n.d.). Estimation of Paddy Rice Planted Area Using Synthetic Aperture Radar - Development of Software "INAHOR". https://www.eorc.jaxa.jp/SAFE/prototyping/accomplishments/inahor/?utm_source.

Stehman, S. 2009. Sampling Design for Accuracy Assessment of Land Cover. *International Journal of Remote Sensing*. 30. pp. 5243–5272.

Takeuchi, W. and Y. Yasuoka. 2004. Development of Normalized Vegetation, Soil and Water Indices Derived from Satellite Remote Sensing Data. *Journal of the Japan Society of Photogrammetry and Remote Sensing (JSPRS)*. 43 (6). pp. 7–19.

Wee, S. 2024. *Improvement and Utilization of Farm Map (Agricultural Digital Map)*. https://www.fao.org/docs/statisticslibraries/apcas/apcas_24_a2-2_improvement-and-utilization-of-farm-map-(agricultural-digital-map)-(korea)-pptx.pdf?sfvrsn=2ffa94cb_1 (accessed March 2025).

World Bank. 2025. Survey Solutions CAPI/CAWI platform: Release 25.01.2. Washington DC, The World Bank.

WorldCereal. 2025. https://esa-worldcereal.org/en (accessed 11 June 2025).

Zhang, H. et al. 2023. Comparing Three Methods of Selecting Training Samples in Supervised Classification of Multispectral Remote Sensing Images. *Sensors*. 23 (20). 8530.

www.ingramcontent.com/pod-product-compliance
Lightning Source LLC
Chambersburg PA
CBHW041759220926
51847CB00014B/262
9789292774707